Measurement
Using the
New Rules of Measurement

Measurement Using the New Rules of Measurement

Sean D.C. Ostrowski

WILEY Blackwell

This edition first published 2013 © 2013 by John Wiley & Sons, Ltd

Registered office: John Wiley & Sons, Ltd, The Atrium, Southern Gate, Chichester, West Sussex, PO19 8SQ, UK

Editorial offices: 9600 Garsington Road, Oxford, OX4 2DQ, UK
The Atrium, Southern Gate, Chichester, West Sussex, PO19 8SQ, UK
2121 State Avenue, Ames, Iowa 50014-8300, USA

For details of our global editorial offices, for customer services and for information about how to apply for permission to reuse the copyright material in this book please see our website at www.wiley.com/wiley-blackwell.

Library of Congress Cataloging-in-Publication Data
Ostrowski, Sean D.C.
Measurement using the new rules of measurement / Sean Ostrowski. – First edition.
pages cm
Includes bibliographical references and index.
ISBN 978-1-118-33301-3 (pbk.)
1. Building–Estimates. I. Title.
TH435.O846 2013
692'.5–dc23
2013027316

A catalogue record for this book is available from the British Library.

Wiley also publishes its books in a variety of electronic formats. Some content that appears in print may not be available in electronic books.

Cover image courtesy of Shutterstock
Cover design by Steve Thompson

Set in 10/12.5 pt Minion by Toppan Best-set Premedia Limited
Printed and bound in Malaysia by Vivar Printing Sdn Bhd

[1   2013]

Contents

Appendices giving answers to self-assessment exercises are on the website
www.wiley.com/go/ostrowski/measurement

The book's companion website is at

www.wiley.com/go/ostrowski/measurement

You will find here freely downloadable support material

The author's website is at

http://ostrowskiquantities.com

Preface

'*Measurement using the New Rules of Measurement*' is intended to provide some guidance on how to measure quantities in accordance with NRM 2 using comprehensive and detailed examples of most of the sections in the NRM 2. The text and examples have been drawn from my professional and academic practice as a chartered quantity surveyor and lecturer.

The publication by the RICS of the suite of New Rules of Measurement (NRM) provides a prescriptive approach to the measurement of quantities throughout the construction process. The second of these, NRM 2, provides new methods of measurement for work in separate trades. As is often the case with innovations, some guidance can be useful and a commentary with examples and exercises on how to use NRM 2 is therefore appropriate.

The publication of the RICS Black Book guidance notes on acceleration and damages for delay to completion provides best practice to the quantity surveying profession and could, perhaps, indicate further developments in standard methods of measurement in an area very much in need of an accurate and consistent approach.

For students and practitioners the acquisition of technical competencies is by practice. A textbook can only provide an introduction. For this reason each chapter has a step-by-step worked example that can be followed and an opportunity to practise with an exercise on each topic. Development can be monitored by using the self-assessment marking sheets that are also provided.

A work like this will contain errors and they are entirely my own responsibility. I would be grateful for your assistance if you would be kind enough to please point out these errors and I will correct them at the first opportunity. I am also aware that some of the opinions in this volume will not be shared by all. I welcome your opinions which I will carefully consider.

Sean D.C. Ostrowski
Summer 2012

Acknowledgements

The author and the publisher would like to thank the following individuals and organizations for their kind permission to use the material as described below:

Barry Symonds for allowing me to use the drawings and my lecture notes.

Andrew Bateson at Cadwyn Housing Association for the use of the partition, flat roof and external works drawings.

Paul Hunt at St George Central London Ltd for the use of the services drawings and the steelwork pergola.

Balkees Noor Mohamed at the College of Estate Management for the use of the drainage drawings.

Tim Cook at Causeway Technologies Limited for the use of the CATO examples.

Palgrave for the use of the windows and door openings drawings.

Phil Young at CEREA for the use of the curtain walling drawings.

Wiley-Blackwell for the use of the sloping site, underpinning and RC frame drawings.

Many people have read and commented on parts of the manuscript and they have my sincere thanks and gratitude:

My wife, Sally, for her assistance with proof reading.

Keith Tweedy for his careful appraisal and discussions on the NRMs.

David Benge for our initial discussions on NRM 1.

Professor Allan Ashworth for his encouragement and tolerance.

David Hockley, RIBA, and Matthew Boughton who despite busy workloads prepared many of the drawings.

Ben Taylor.

David Quarmby of Glamorgan University.

Simon Pope at Causeway Technologies Limited.

The students of several institutions whom I have been privileged to teach who diligently and sometimes gleefully but always in good humour pointed out everything that they could find that was wrong or inconsistent and were kind enough to tell me what they found difficult about what they were being taught.

My publishers Wiley Blackwell for their encouragement and comments for which I am profoundly grateful.

List of Drawings

Practical applications	Self-assessment exercises
	Protocol photographs 1–4
	SDCO/2/E3/1 Corners
SDCO/2/4/1 Substructure plan	SDCO/2/E4/1 Trench
SDCO/2/4/2 Substructure section	
SDCO/2/5/1 Basement plan	SDCO/2/E5/1 Basement plan and section
SDCO/2/5/2 Basement section	
SDCO/2/6/1 Sloping site plan	SDCO/2/E6/1 Sloping site plan
SDCO/2/6/2 Sloping site sections	SDCO/2/E6/2 Sloping site section 1
	SDCO/2/E6/3 Sloping site section 2
SDCO/2/7/1 Underpinning plan	SDCO/2/E7/1 Underpinning
SDCO/2/7/2 Underpinning section	
SDCO/2/8/1 Reinforced concrete frame plan and details	SDCO/2/E8/1 Formwork
SDCO/2/8/2 Reinforced concrete frame section	
SDCO/2/9/1 Brickwork plan	SDCO/2/E9/1 Plan
SDCO/2/9/2 Brickwork elevations	SDCO/2/E9/2 Casement window
SDCO/2/9/3 Brickwork details 1	SDCO/2/E9/3 Circular window
SDCO/2/9/4 Brickwork details 2	SDCO/2/E9/4 External door
SDCO/2/9/5 Brickwork details 3	
SDCO/2/10/1 Windows elevation, section and details	
SDCO/2/10/2 Glazing	
SDCO/2/11/1 Flat roof plan	As Chapter 11
SDCO/2/11/2 Flat roof section and details	
SDCO/2/12/1 Pitched roof plan and section	SDCO/2/E12/1 Pitched roof
SDCO/2/13/1 Steelwork plan and sections	SDCO/2/E13/1 Pergola
SDCO/2/14/1 Partitions first floor plan	SDCO/2/E14/1 Partitions ground floor plan

Practical applications	Self-assessment exercises
SDCO/2/14/2 Partitions section	
SDCO/2/15/1 Curtain walling plan and details	SDCO/2/E15/1 Curtain walling
SDCO/2/15/2 Curtain walling details	
SDCO/2/16/1 Finishes plan	As Chapter 16
SDCO/2/16/2 Finishes elevation	
SDCO/2/17/1 Drainage plan	SDCO/2/E17/1 Drainage plan
SDCO/2/18/1 Hot and cold water plan	SDCO/2/E18/1 Plumbing plan
SDCO/2/18/2 Soil and waste pipework	SDCO/2/E18/2 Plumbing isometric
SDCO/2/18/3 Schematic cold water	
SDCO/2/19/1 Air conditioning plan	SDCO/2/E219/1 Ventilation
SDCO/2/19/2 Air conditioning sections	
SDCO/2/19/3 Air conditioning details	
SDCO/2/20/1 Small power plan	SDCO/2/20/E1 Lighting layout
SDCO/2/20/2 Power symbols	SDCO/2/20/E2 Lighting symbols
	SDCO/2/20/E3 Cable schematic
SDCO/2/21/1 Retaining wall plan 1	As Chapter 21
SDCO/2/21/2 Retaining wall plan 2	
SDCO/2/21/3 Retaining wall elevation and section	

List of Tables

List of Diagrams

Glossary of Terms and Abbreviations

abd	as before described
b & s	bed and surround
b.i.	built in
BIM	building information modelling
bkwk	brickwork
blwk	blockwork
BMS	building management system
BQs	Bills of Quantities
BS	British Standard
BWIC	builders' work in connection
CATO	computer aided taking off
c/cs	centres
CCTV	closed circuit television
CL	cover level
c/m	cement mortar
CP	cost plan
CWS	cold water system
Ddt	deduct
DPC	damp proof course
DPM	damp proof membrane
EGL	existing ground level
EML	expanded metal lathing
EO	extra over
FAI	fresh air inlet
FCU	fan coil unit
f/f	fair faced
FL	formation level
gl	ground level
g/m	gauged mortar
gwl	ground water level
h/c	hardcore
hw	hardwood
HWS	hot water system
IL, iv	invert level
IT	invert trap
iv	invert level
lm	linear metre
M&E	mechanical and electrical
MH, m/h	manhole

ms	measured separately
ne	not exceeding
NRM	New Rules of Measurement
O/A	overall
PCC	precast concrete
PIB	polyisobutylene
PPP	public–private partnership
QS	quantity surveyor
RC	reinforced concrete
RE	rodding access
RIBA	Royal Institute of British Architects
RICS	Royal Institution of Chartered Surveyors
RL	reduced level
R&S	render and set
RWP	rainwater pipe
SMM	standard method of measurement
s/s	stainless steel
sw	softwood
T & C	test and commission
th	thick
WHB	wash hand basin
WBP	weather and boil proof

1 Introduction

1.1 Introduction
- Contents
- Practical examples and self-assessment exercises
- Royal Institution of Chartered Surveyors (RICS) competence levels
- Companion websites

1.2 Standard methods of measurement
- Elemental measurement
- Trade measurement
- Compatibility
- The relevance of Bills of Quantities
- The advantages of BQs
- The disadvantages of BQs

1.3 Contract documentation
- Components
- Documentary discipline

1.1 INTRODUCTION

Contents

The contents include an introduction to most of the sections in *New Rules of Measurement NRM 2* intended for use by students and practitioners as follows:

1.	Introduction
2.	The Basics
3.	Corners
4–21.	Individual trades
22.	Preliminaries
23.	Computer aided taking off
24.	Preparation of Bills of Quantities (BQs)

Measurement Using the New Rules of Measurement, First Edition. Sean D.C. Ostrowski.
© 2013 John Wiley & Sons, Ltd. Published 2013 by John Wiley & Sons, Ltd.

Also included is:

- A detailed worked example of the practical application in each chapter
- A comprehensive exercise for practice at the end of each chapter
- The detailed answer with the calculations
- A self-assessment marking sheet to provide an indication of the standard achieved in technical and managerial competence and cognitive development
- A companion website with animated Powerpoint presentations of the chapters to provide further assistance and opportunities to ask real time questions

The first edition of SSM 7 was published in 1988. There are several developments that make the introduction of a new method of measurement appropriate. The first is the decline of trade crafts. Most site work is semi-skilled at best and does not require the complex labour skills that were necessary in the past. NRM 2 has reduced the extent of the 'labour' items that are required to be measured. Another development is the technological development of off site fabrication with site erection and the consequent decline in the 'wet' trades like plastering. Partitioning and curtain walling have become a large proportion of construction work and are now representedin specific sections in the NRM 2 as Section 20 Proprietary linings and partitions and Section 21 Cladding and covering. Finally many client bodies have become active partners in the construction process and these knowledgeable employers administer the contracts from beginning to end. They have become aware of the inconsistencies in financial reporting throughout the construction cycle. These developments make a new suite of standard methods of measurement appropriate.

The New Rules of Measurement (NRM) provide an accurate and consistent approach through the full life cycle of the building at each stage of development: the estimate; the cost plans; the work packages and BQs and finally the whole life costing maintenance programmes. The intention is to provide an audit trail of the quantities and prices from the beginning to the end of the life cycle of the building. There is an expectation that the NRM will have a wide appeal with an opportunity for countries around the world to adopt a common set of rules for the measurement of building works.

The Royal Institution of Chartered Surveyors (RICS) provides the following documents that will help to provide comprehensive, accurate and consistent financial reporting. They are:

- *The RICS Code of Measuring Practice*, 6th edition, 2007
- *The RICS new rules of measurement NRM 1: Order of cost estimating and cost planning for capital works*, 2nd edition, April 2012
- *The RICS new rules of measurement NRM 2: Detailed measurement for building works.*, April 2012
- *The RICS new rules of measurement NRM 3: Maintenance and operations cost estimating, planning and procurement*. To be published in 2013

Practical examples and self-assessment exercises

The pedogogy of quantity surveying is primarily concerned with the nature of the knowledge that is being learned. Technical competencies such as estimating, measurement and contract administration are procedural knowledge. Practice is the most useful method of teaching procedural knowledge (Gagne, 2002) and teaching in the form of telling or demonstrating to the student how to do it is not effective (Wood, 2001), A technical competence like measurement is acquired by practice. The use of textbooks, lectures or demonstrations only provides an introduction to how to measure. The most effective way of learning how to measure is to practise and to receive prompt answers to questions as they arise. However before the practice can take place it is helpful to examine examples to see the process that is required and how to set out the work. An examination of the practical examples in each chapter will provide the information necessary to carry out the self-assessment exercise which follows after the practical example.

The practical examples and self-assessment exercises are set out on traditional rulings, double dimension paper. Although much work can now be done on spreadsheets and software, the need to understand the construction technology, the use of side casts and the conversion calculations to enable the measurement to be compliant with the NRM are all easier to understand if set out on double dimension paper. When the competence levels have reached Level 2, an ability to carry out the work comprehensively and accurately without supervision, then the software programmes can be introduced. Software measurement packages require a significant amount of practice before they can be used effectively. Proficiency in the software is best acquired after expertise in measurement has been attained.

RICS competence levels

The Royal Institution of Chartered Surveyors (RICS) Assessment of Professional Competence (APC) comprises the demonstrable acquisition of a series of competencies after a period of time in the profession. This includes the provision of a diary showing a structured training programme and a final assessment interview. Two of the competencies are 'Design Economics and Cost Planning' and 'Quantification and Costing of Construction Works'. The first concerns estimating and cost planning and the second is measurement. They are core technical competencies that are mandatory for the successful completion of the APC. Each competence has three levels. Level 1 is knowledge about the subject. This is the provision of propositional knowledge about the subject. Level 2 is being able to apply the knowledge. This is the provision of procedural knowledge of actually being able to undertake the competence to the level of skill that is both comprehensive and accurate. Level 3 is being able to discriminate the quality of the work and advise the client and will only be available after some time in practice, which is part of the RICS APC programme. In this textbook the provision of worked examples can be followed and replicated and practical exercises at the end of each chapter provide practice at Level 2. The marking scheme provides some level of discrimination on the quality of the work that has been practised. Knowing how to use these NRMs is therefore a necessary step on the route to Level 3 and a professional qualification as a chartered quantity surveyor.

Companion websites

Many students, particularly at the outset, find the subject difficult. The printed word in the form of a textbook has a limited usefulness in providing the appropriate teaching for these technical competencies. The most effective method of acquiring expertise in these disciplines is by practice and the contemporaneous answers to questions as they arise. To provide further assistance there are dedicated websites at http://ostrowskiquantities.com with full A4 formats of the practical examples and exercises, animated Powerpoint presentations of each trade described in the text and opportunities to ask real time questions, attend workshops, seminars, conferences and courses, and at Wiley Blackwell (http://www.wiley.com/go/ostrowski/measurement). It is hoped that the provision of these additional facilities will go some way towards expanding the opportunities for practice in a more useful way than using the printed word alone.

The RICS website includes the NRMs free of charge for members and they are also available on their subscription information service, ISurv. Most practices, contractors and universities are subscribers. This means that a screen based version is available to most individuals free of charge and hard copies can be obtained for the cost of the printing.

1.2   STANDARD METHODS OF MEASUREMENT

There are currently several different methods of measurement for building works published by the RICS as follows:

- *The RICS Code of Measuring Practice.* This sets out how to measure floor areas.
- *The RICS new rules of measurement NRM 1: Order of cost estimating and cost planning for capital works.* This provides a method of measurement for quantities on an elemental basis for estimates and a different method of measurement for cost plans.
- *The RICS new rules of measurement NRM 2: Detailed measurement for building works.* This measures quantities on a trade basis.
- *The RICS new rules of measurement NRM 3: Maintenance and operations cost estimating, planning and procurement.* This measures quantities on an elemental basis as NRM 1.

The introduction of the full suite of NRMs means that there are accurate forms of measurement that will need to be prepared to reflect the design and specification at the end of significant stages of the project.

There are other formats for standard methods of measurement for specific purposes, and many countries have their own formats. Examples include the Civil Engineering Standard Method of Measurement (CESMM) and the Principles of Measurement International (POMI).

Elemental measurement

The first method of measuring quantities is by using the elemental method of measurement in NRM 1 (*Order of cost estimating and cost planning for capital works*). Each element of the building is separately measured, e.g. a reinforced concrete roof will be a separate section comprising several trades, *viz.* concrete, formwork, reinforcement, screed, asphalt, metalwork, balustrading, etc.

Within this elemental method there is a different method of measurement for the estimate and the cost plan. The estimate uses preparatory design information and uses mainly the superficial floor areas as the basis for measurement. The cost plans use a progressively more developed design and measure quantities using units that are cubic superficial, enumerated and itemised for a larger range of elements.

Trade measurement

The second method of measurement is measuring the work by using the trade method of measurement in NRM 2 (*Detailed measurement for building works*). Each trade is measured separately wherever it occurs in the building using a technically complete design. This enables the efficient collection of trade works into separate Bills of Quantities and ease of pricing by the contractor.

Compatibility

The goal of a strong, seamlessly linked cost control pathway has commenced with the publication of NRM 1 and 2 which provide sets of rules that are accurate and consistent. It can also be seen that NRM 1 and 2 provide more than one set of rules which are alternative and overlapping methods for the measurement of quantities. The progressive measurement and pricing stages for estimates, cost plans and trades provide a structured cost management framework that is more detailed and accurate at each stage. However standardisation, accuracy and consistency will be lost if the prices in estimates and cost plans which have been prepared on an elemental basis using NRM 1 are not compatible with the tendered prices which have been prepared on a trade basis using NRM 2. The audit trail will be disconnected and the transparency that the client requires may not be possible. The measurement and pricing in the cost plans cannot be compared with the measurement and pricing in the trade Bills of Quantities (BQs). The client may consider that going out to tender on the elemental measurements included in the cost plans will provide adequate early stage prices. However to provide financial security for the client the measurement of trade BQs using NRM 2 can be priced by the contractor to provide a fully quantified schedule of rates. They are a complete and detailed analysis of the measurement and pricing of all the works and in this way they protect all parties. In this way the perception that the BQs are a barrier to collaboration amongst the stakeholders is removed.

The advantage of elemental measurement is the relevance of the costs to a particular part of the building. The advantage of trade measurement is that this is the basis of the pricing for the contractors and subcontractors who construct the work. Weights and volumes will remain the basis of pricing substantial parts of all construction work and therefore will remain the basic form of measurement. However it is possible to combine the advantages of both NRM 1 and NRM 2 into a single structured set of rules. The appropriate parts of the trade BQs can be allocated to the appropriate elements.

The relevance of Bills of Quantities

The measurement of the quantities of work in a construction project is an important part of establishing what the cost will be. These quantities were originally prepared by each contractor and then by a separate organisation who sold the quantities to the contractors. Quantity surveying (QS) practices then developed and they provided the BQs to the client as part of the financial control. The client then sent the BQs to the contractors to provide a tender for the work. Recent developments have seen work packages prepared by the consultants which comprise the drawings and specifications but do not include BQs. The package contractor prepares his own BQs and they may not form part of the contract documentation. The work is still measured and the measurement is often carried out by the same QS practices. The professional practices now prepare the BQs for the subcontractors. This is because a successful contract requires an accurate price and this requires accurate quantities. It can be seen that BQs will continue to be needed. The recent launch of the NRM suite of standard methods of measurements has brought into focus two significant developments. The first is the perceived understanding that BQs act as a barrier between the Client and the professionals. The second is the need to standardise methods of measurement throughout the whole life of the project, from inception to demolition. The merits of an accurate set of quantities from a third party remain considerable; at the same time there are also some disadvantages. They are both rehearsed below.

The advantages of BQs

BQs provide a comprehensive and accurate measurement of the total work necessary to be completed. The pricing of such a document can provide a comprehensive and accurate bid for the work. They provide an excellent method of obtaining competitive prices and they are an excellent vehicle for comparing the tenders. BQs are an excellent check of buildability because the design is examined and measured in such detail. During construction they can be used to measure the amount of work completed as the basis for interim certificates. Variations can be easily priced using the BQs as a schedule of rates. The final account can use the BQs as the basis for the final certificate.

The disadvantages of BQs

The fundamental problem with BQs is that they form a barrier to understanding by the design team. They do not form an integral and seamless transition from the estimate to completion that can be understood by all parties. The rules for the preparation of BQs are not the same as the rules for forming the estimate or cost plan. They are an excellent way of getting the cheapest price and for post-contract financial control but they can only be used and scrutinised by the QS team who have the specialist knowledge required to interrogate the figures. The solution is to amend the structure of trade BQs into a format compatible with that of the estimates and cost plans. They require a large amount of information from the design team and also take time to prepare. Because they are comprehensive and accurate they are expensive to prepare. These costs are incurred before the work has started on site and constitute a significant drain on the developer's cash flow. Much work is now let as work packages which require smaller and discrete amounts of information and measurement. A large proportion of work is now built using design and build contracts where the risk for accurate design and pricing lies with the contractor. The need for comprehensive and accurate BQs is not so important with these contracts.

1.3   CONTRACT DOCUMENTATION

Components

The BQs form part of the contract documentation. The components are as follows:

- Contract
- Drawings
- Specification
- BQs
- Tender
- Statutory requirements
- Employers' requirements
- Site requirements
- Contractor requirements

Documentary discipline

The documentation is extensive and comprehensive. Such a complex array of documentation requires rigorous discipline to ensure that the information remains relevant and accessible. The use of electronic storage and cut and paste from one contract to the next often leads to duplication and error. The quality of the documentation is often compromised and is superseded by a vast array of disorded documents in the hope that the required clauses will be included somewhere in the documents.

2 A Practical Introduction to Measurement

2.1 A practical introduction to measurement
- Measurement protocols
- Scales
- Accuracy
- Builders' quantities
- Symbols
- Query sheets/to take lists/marked up drawings
- Revisions
- Information

2.2 Measurement procedure
- Technical competence
- Procedure
- Levels
- Compound items

2.3 Self-assessment exercise: Protocols

2.1 A PRACTICAL INTRODUCTION TO MEASUREMENT

Measurement protocols

The following example (see Table 2.1) sets out a typical 'taking-off' sheet. Each item is numbered and explained as follows:

1. 'Rulings' is the collective name for the various kinds of layouts used by the QS and refers to the type of vertical lines used on the paper. This example is double dimension paper which has two columns. Both columns can be used.
2. Column 2.
3. Every page has 'headers' to identify the project and 'footers' for page numbers.
4. The dimension column sets out the dimensions in metres to two decimal points.
5. The 'timesing' column provides any multiples that are necessary

Measurement Using the New Rules of Measurement, First Edition. Sean D.C. Ostrowski.
© 2013 John Wiley & Sons, Ltd. Published 2013 by John Wiley & Sons, Ltd.

6. The 'squaring' column gives the product and sum of each of the dimensions. It is double underlined as it is an item that will go into the BQ.
7. This column provides the description of the item using the appropriate elements and levels set out in the NRM. They do have to be grammatically correct and do not need a full stop.
8. This is the NRM reference.
9. Calculations that may be necessary to provide a dimension are 'side-casts' (also called 'waste calculations'). They are dimensions shown on the drawings which are always shown in millimetres.
10. Cubic dimensions, length × breadth × depth.
11. Superficial dimensions, width first multiplied by height.
12. Linear dimensions.
13. Dotting on allows further multiples of dimensions without repeating the dimension.
14. Enumerated dimensions.
15. Items that are not measured.
16. Brackets (thick black lines in this volume) to contain sets of dimensions or descriptions.
17. A 'rogue' item that is significant enough to be measured but does not fit into the NRM format. NRM 2 items 3.3.5.1–2, pp.48–9 explain that the measurement should be similar. An example could be a unique set of labours specific to that particular contract only.

Table 2.1 Measurement protocols.

	1. Column 1			2. Column 2		

3 Excavating and filling

Column 1 — columns: 5, 4, 6, 7

5	4	6	7

9

10000
5000
15000

10
15.00
1.50
.75

| Excavation, foundation, ne 2m deep **8** [5.6.2.1

11
10.00
5.00

Walls, ½ brick thick, skins of hollow walls, g/m 1:1:6, stretcher bond
[14.1.1.1

12

4/2/ 2.00 16.00 | Architrave, softwood,
4/ 1.00 4.00 | 50 x 20, moulded x 3
13 3.2/ 3.00 | [22.2.1.2

15.00
35.00

&

16

Door stops, 25 x 12, pelleted
[24.11.1.1

Column 2 — columns: 5, 4, 6, 7

5	4	6	7

14

6/ 1

Access panels, 300 x 300
[30.7.1

15

Item

Testing and commissioning, plumbing
[33.10.1

17

10.00
10.00

Rogue item eg Decorative pargetting to the surface of plastered finishes

The abbreviations 'abd' (as before described), 'ditto', and 'do' can be used to indicate repeated descriptions.

3

4

Scales

Drawings can come from a variety of sources, free form sketches, hand drawn plans, elevation and sections on paper and several software packages that make two and three dimensional drawings. Typical scales are:

- 1:1250   Small scale drawings to show large areas eg landscaping
- 1:100   Design drawings for layouts and elevations eg plans
- 1:50   Production drawings to show the construction eg sections
- 1:5   Large scale drawings to show details eg interfaces

NRM 2 states at clause 2.14.4.1 *'Drawings shall be to a suitable scale'.* The extensive electronic transmission between terminals and software packages and printing on different sizes of paper will often make these scales unusable. This reduction and enlargement of drawings is common between computer terminals and printers and care needs to be taken to ensure that the scales shown on the drawings are not used to extrapolate dimensions that are not shown as figured dimensions. The scale of the drawings is one of the first matters that should be checked.

For our working examples and exercises throughout this book the drawings often distort the dimensions to such an extent that they cannot be scaled. Figured dimensions on the drawings are the only dimensions that should be used. The improvement that technology has provided in the rapid electronic exchange of drawings has been offset by the problems of changes in scale with enlargements and reductions of the printed drawings. Dimensions that are scaled or extrapolated should be confirmed before being used in published documents. These dimensions can be included on a query sheet similar to the example provided in Chapter 3 (Table 3.1) and sent to the Architect for confirmation. All dimensions should be figured dimensions provided by the designers.

Accuracy

One of the major anxieties that students encounter is how accurate should these measurements be? The desire for accuracy to the nearest millimetre is often the first step of the diligent student. However, this is not necessary. Guidelines set out in NRM 2, page 46, item 3.3.2(d), state that dimensions used in calculating quantities shall be taken to the nearest 10mm and at 3.3.2(e) that quantities shall be to the nearest whole number. This can be illustrated by measurement in cubic metres.

A dimension of $3460 \times 3460 \times 3460 = 41.42m^3$. This is $41m^3$ to the nearest whole number
A dimension of $3500 \times 3500 \times 3500 = 42.88m^3$. This is $43m^3$ to the nearest whole number
A dimension of $3525 \times 3525 \times 3525 = 43.80m^3$. This is $44m^3$ to the nearest whole number

This demonstrates that a range of 65mm from 3460 to 3525 only increases the cubic quantity by $3m^3$.

Although the popularity of BQs fluctuates, the need for accurate quantities continues. Currently most work is tendered on a trade basis, in accordance with the nature of the construction industry. Part of the tendering process is the measurement, although it may not be formally prepared or included in the contract documentation which may be limited to 'design and build' or 'drawings and specification' or indicative work packages. The prudent subcontractor will measure the work and the measurement is often undertaken by the same professional QS practices that prepare bills of quantities for the client. The cost is included in the tender not in the fees.

Builders' quantities

Many contractors and subcontractors develop their own method of measurement to suit their own areas of expertise. These are 'builders' quants'. They are fast and effective and provide the contractor's estimator with the essential information to provide a price. For example, plasterers and painters may count the number of rooms to ascertain the labour content and ignore the materials content. They know how long it takes to do a room and use these specialist labour constants to provide a reasonably accurate estimate of the cost. Curtain walling contractors will separate materials and labour. The materials cost will come from the specialist manufacturer and the labour costs will be provided by the labour-only installers. Publishing these documents is problematic for several reasons. They are unique to each contractor and usually emphasise the particular areas of cost that are considered to be significant. They are closer to being an estimate than a detailed and accurate schedule of quantities. They are not in accordance with the RICS guidelines, NRM 2 p.1, which indicates that using NRM 2 is recommended good practice. Using 'builders' quants' for valuations and variations could lead to inaccuracies and could only be possible by an amendment to the Form of Contract and agreed by both parties.

Symbols

NRM 2, Section 1.6 Symbols, abbreviations and definitions, lists a specific set of symbols which are to be used. Take as an example the symbol 'No.' to indicate an enumerated item like 5 No. manholes. This may be your normal method for describing enumerated items. However, the NRM says that this should be shown as 5 nr manholes. There is no capital or full stop. The point as a decimal marker (00.11) and the comma as a thousand separator (10,000.00) are almost universally used in construction. However, the manufactured products used in construction often use a different convention. To abandon the use of the normal construction conventions would render the quantities in any BQs confusing and misleading. The Code of Measuring Practice provides a timely warning. '*The British Standard BS 8888: 2006 Technical Product Specification (for defining, specifying, and graphically representing products) recommends the inclusion of a comma rather than a point as a decimal marker, and a space instead of a comma as a thousand separator. While the convention has not been adopted in this Code, users should take care to ensure that this does not conflict with client requirements*' (The Code of Measuring Practice, 6th edition 2007, p. 5).

Query sheets/to take lists/marked up drawings

The expertise that quantity surveyors have acquired enable them to examine the drawings and highlight areas that need clarification, additional information or possible corrections. Where there are queries on the drawings they should be scheduled on a query sheet. By providing the query sheets to the design team they are assisting the design process to the benefit of all concerned.

A 'to take' list to act as an *aide-mémoire* to ensure that nothing is forgotten will be of assistance in ensuring that each item is assigned to the correct trade and that rogue items are also measured to reflect the full extent of the work.

'Marked up' drawings to demonstrate that all the work has been measured are a simple and effective way to ensure that nothing has been left out.

Revisions

The electronic transmission of revised drawings without a revision being timed or dated and partial alterations to parts of existing drawings mean that the particular drawing or detail that is being used should be carefully recorded on the drawing schedule, prepared by the quantity surveyor, and noted on the dimensions.

Information

Extensive schedules of information requirements are set out in NRM 1 for estimates and cost plans. NRM 2 also provides information requirements for measurement (p. 30, Section 2.14). The provision of information in a timely manner remains the biggest challenge in the measurement process. It is hoped that the formalisation of these new schedules will assist all parties.

2.2   MEASUREMENT PROCEDURE

Technical competence

Measurement is a technical competence using procedural knowledge. This is the knowledge of how to do things, how to carry out functional activities. Examples include riding a bicycle, playing the piano and bricklaying. The most important requirement of this work is to practise the procedure to enable improvement and expertise. Increased competence with the procedure comes with practice. The practice includes a procedure or process similar to a recipe for making a cake or addressing a golf ball. The elements are precisely defined in terms of quality and quantity and they occur in a predetermined sequence. It does not require any innovative thinking and it is not necessary to know the solution before the process commences. A successful outcome is simply a matter of following the process. The more it is practised the higher the level of competence that is acquired.

Procedure

Problems are often manifest in the form of the question 'What do I do next?' A procedure is set out below to address this question.

- The front page is the title page. It describes the project, the NRM section, the drawings, the specification, the author and the date.
- Each page is 'topped and tailed' with the reference at the top and pagination at the bottom.
- Look at the drawings to ascertain that all the required information is available.

- Prepare a 'to take' list of all the items that need to be measured, which includes all the NRM items and any rogue items on the drawings that should also be measured.
- Prepare a query list to provide outstanding information.
- Calculate the 'side-casts' that are likely to be used, e.g. centre lines, floor to ceiling heights, room dimensions.
- Measure the work progressively using the NRM as a guide. This can be:

  All the items in the sequence of construction, or

  All the items in each trade, or

  All the items on each drawing.
- Clear each item from the 'to take' list.
- Mark up the drawings so that all the items are seen to have been measured.
- Prepare a list of outstanding items to be measured when clarification is provided.

Levels

NRM 2 uses a series of levels to describe and measure the work in increasing detail. The more complex the description the more levels will be necessary. In this example – the measurement of reinforced concrete columns for the reinforced concrete (Table 2.2) – all the levels are populated but in the formwork and the reinforcement one of the levels is not used. Levels 1–3 comprise specification information and are often combined into one description. Although the levels are useful they can be amalgamated for manual taking off. However, for computerised software packages they depend entirely on the mechanistic implementation of each of the levels to enable the measurement to be completed.

Table 2.2 Levels for the measurement of reinforced concrete columns.

LEVEL		NRM 2						
1	SECTION	11	In-situ concrete works					
2				In-situ concrete		Formwork		Reinforcement
3				Reinforced in-situ concrete		Plain formwork		Nil
4	SUBSECTION		11.5	Vertical work	11.20	Sides of isolated columns, (nr stated)	11.33	Mild steel bars
	UNIT			m³		m²		t
	NRM 2 LEVELS							
5	LEVEL 1		11.5.1	<300 thick	11.20.1	Regular: shape stated	11.33.1	Nominal size
6	LEVEL 2		11.5.1.1	In structures (Columns)	11.20.1.1	Propping <3m high	11.33.1.1	Straight
7	LEVEL 3		11.5.1.1.2	Reinforced >5%		Nil	11.3.1.1.1	Bars exceeding 12m long

A direct comparison with NRM 1 is not possible but an approximate correlation is set out in Table 2.3.

Table 2.3 A comparison of levels between NRM 1 and NRM 2.

	LEVELS							
	1	**2**	**3**	**4**	**UNIT**	**5**	**6**	**7**
NRM 1 Cost plans	2 Group element Superstructure	2.1 Element Frame	2.1.4 Sub-element Concrete frame	2.1.4.1 Component Columns	m	2.1.4.1.C2 Subcomponents Column grid size		
NRM 2	11 In-situ concrete works	In-situ concrete	Reinforced In-situ concrete	11.4 Vertical work	m³	11.5.1 Level 1 <300th	11.5.1.1 Level 2 In structures, columns	11.5.1.1.2 Level 3 Reinforced >5%

Compound items

NRM2 concentrates the measurement on the items that carry the most costs. It provides a simpler method of measurement. This means that there are fewer items to measure. The remainder the 'labour' items – are not measured, e.g. working space (Level 3 Item 3, NRM p. 132), and level and compact the surface of the excavation to receive the foundations (Level 3 Item 5, NRM p. 132) is no longer measured. The intent is that these low-value labour items are incorporated into the pricing of the work. However, these small-value items still have to be carried out and the contractor's estimator still has to include the cost of the work in the tender price. This requires the build up of a compound item which includes more than one unit of measurement, each one of which has a separate price, e.g. the disposal of excavated material off site is no longer measured and is an Item (NRM 2, Item 5.9). This work requires the quantities from measurement of the excavation, the measurement of any working space that may be required including its disposal and backfill, the bulking factor for the transportation off site, the staging of the materials that may be required, the landfill tax applicable to each load, and the compacting of the surface of the excavation to receive the fill. To provide a comprehensive rate a compound item is necessary as in Table 2.4.

Table 2.4 Compound item.

Excavation, bulk excavation, 2–4m deep		$48m^3$@ £3/m^3	144.00
Working space: excavation		$10m^3$@£25/m^2	250.00
Working space: backfill		$10m^3$@£50/m^2	500.00
Bulking	$48 + 10 = 58m^3 + 25\% = 73m^3$		
Disposal off site		$73m^3$@ £75 /m^3	5,475.00
Staging	Say	$50m^3$@£1/m^3	50.00
Landfill tax	$48m^3 \times 1400/1000 = 67t$	67t @2.50/t	168.00
Compact surface		$20m^2$ @£2/m^2	40.00
NRM 5.9	**Disposal of excavated material off site**	**Item**	**£ 6,627.00**

These examples indicate that the simplification of the method of measurement does not reduce the amount of measurement. The task of the estimator has to expand from simply pricing the items to adding to the measurement that which is necessary to provide a comprehensive tender price. This means that parts of the measurement work have been transferred from the fees of the professional practice to the overheads of the contractor. The client still has to bear the total cost of the tender. Because part of the measurement now lies with the contractor the risk increases and therefore the tender price is likely to increase. Simplification has come at a price which may not be cost effective.

2.3　SELF-ASSESSMENT EXERCISE: PROTOCOLS

Measure the items included in Appendix 2 using blank double dimension paper in Appendix 1a. Please prepare your own measurement and also prepare a query sheet of problems that you have encountered. Compare your own work with the proposed solution in Appendix 2 (Table E2.1). Self-assess your work on the assessment sheet in Appendix 1b.

To provide further assistance there are dedicated websites at http://ostrowski quantities.com and at Wiley Blackwell (http://www.wiley.com/go/ostrowski/measurement). It is hoped that the provision of this will go some way towards explaining the concepts and principles more clearly than using the printed word alone.

3 Corners

3.1 Measurement information
3.2 Methodology
3.3 Practical application: Corners
3.4 Self-assessment exercise: Internal and external dimensions

3.1   MEASUREMENT INFORMATION

The drawings for this exercise are a series of sketches which appear alongside the calculations in the practical application. The specification used to illustrate how to measure round corners is cavity brickwork. The methodology is relevant to any form of construction. The query sheet is used to confirm dimensions that may have been extrapolated or scaled.

Table 3.1 Query sheet.

QUERY SHEET	
Dimensions	
QUERY (From the QS)	ANSWER (From the Architect/Engineer) (Assumptions are to be confirmed by QS)
1. Thickness of outer skin 2. Width of cavity 3. Thickness of inner skin	Assumed Assumed Assumed

Measurement Using the New Rules of Measurement, First Edition. Sean D.C. Ostrowski.
© 2013 John Wiley & Sons, Ltd. Published 2013 by John Wiley & Sons, Ltd.

3.2   METHODOLOGY

There are a series of calculations, called side casts, that occur on all buildings. The first is the centre line calculation which allows the girth or perimeter of the building to be measured through the centre line of the external wall. The calculation can use internal dimensions, those that measure to the inside face of the building, or external dimensions, those that measure to the outside face of the building. The second is the spread of foundations which measures the differential between the thickness of the foundations and the walls. These first two are examined in this chapter. Others are the formation level to determine the excavation for the foundations which is examined in Chapter 4; the calculations of the structural height of each floor level and the floor to ceiling heights which are examined in Chapter 16; the structural opening size for doors and windows which is examined in Chapter 10; the depth of drainage trenches derived from the datum levels which is examined in Chapter 17. These side casts can be calculated early in the taking-off process and establish significant dimensions which are used often.

Table 3.2 Practical application: Corners.

			Side casts				
			CALCULATIONS				
			Drawings SDCO/2/3/1–5 Sketches				
			Specification N/A				
							The title page includes the trade name, the name of the contract, the full drawing schedule with dated revisions, the dated specification, the name of the measurer and the date.
			SDCO May 2012				*The scale of the drawings is also normally included but in this textbook they have been excluded.*

Table 3.2 Practical application: Corners (*Continued*)

Side casts: centre lines

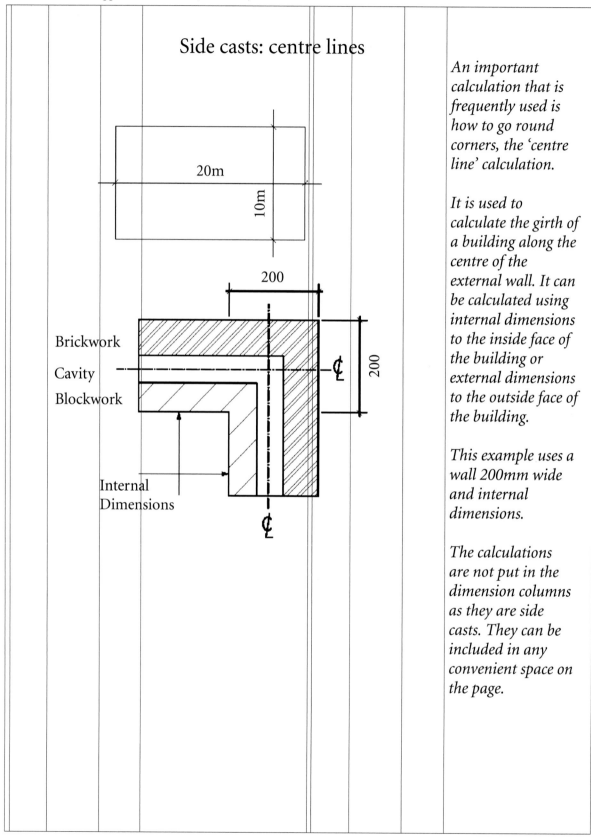

Brickwork

Cavity

Blockwork

Internal
Dimensions

An important calculation that is frequently used is how to go round corners, the 'centre line' calculation.

It is used to calculate the girth of a building along the centre of the external wall. It can be calculated using internal dimensions to the inside face of the building or external dimensions to the outside face of the building.

This example uses a wall 200mm wide and internal dimensions.

The calculations are not put in the dimension columns as they are side casts. They can be included in any convenient space on the page.

Table 3.2 Practical application: Corners (*Continued*)

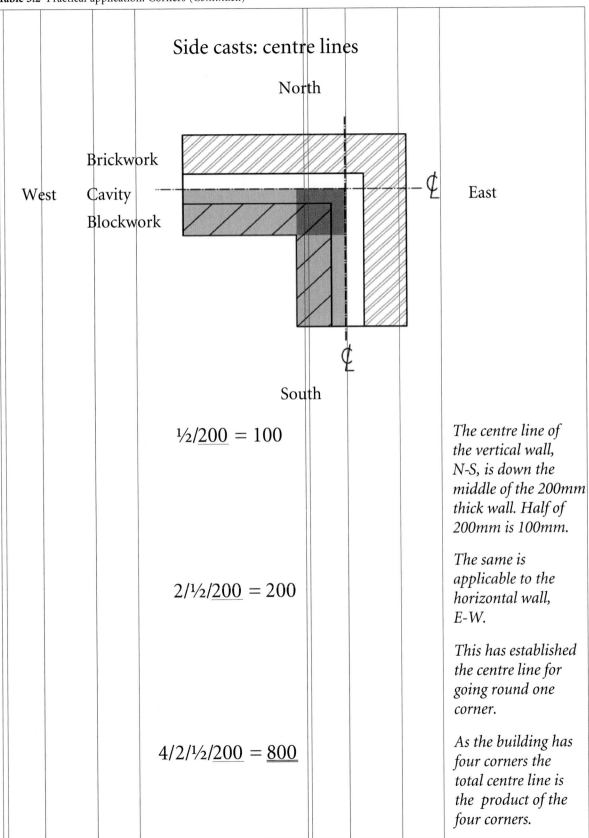

Side casts: centre lines

North

Brickwork

West   Cavity   ₵   East

Blockwork

South

½/200 = 100

2/½/200 = 200

4/2/½/200 = 800

The centre line of the vertical wall, N-S, is down the middle of the 200mm thick wall. Half of 200mm is 100mm.

The same is applicable to the horizontal wall, E-W.

This has established the centre line for going round one corner.

As the building has four corners the total centre line is the product of the four corners.

Table 3.2 Practical application: Corners (*Continued*)

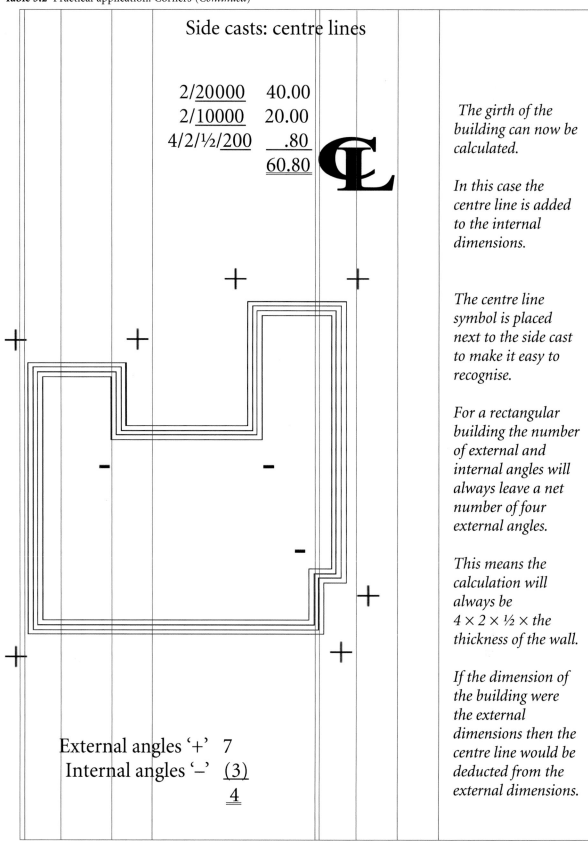

Side casts: centre lines

2/20000	40.00
2/10000	20.00
4/2/½/200	.80
	60.80

℄

External angles '+' 7
Internal angles '−' (3)
4

The girth of the building can now be calculated.

In this case the centre line is added to the internal dimensions.

The centre line symbol is placed next to the side cast to make it easy to recognise.

For a rectangular building the number of external and internal angles will always leave a net number of four external angles.

This means the calculation will always be 4 × 2 × ½ × the thickness of the wall.

If the dimension of the building were the external dimensions then the centre line would be deducted from the external dimensions.

Table 3.2 Practical application: Corners (*Continued*)

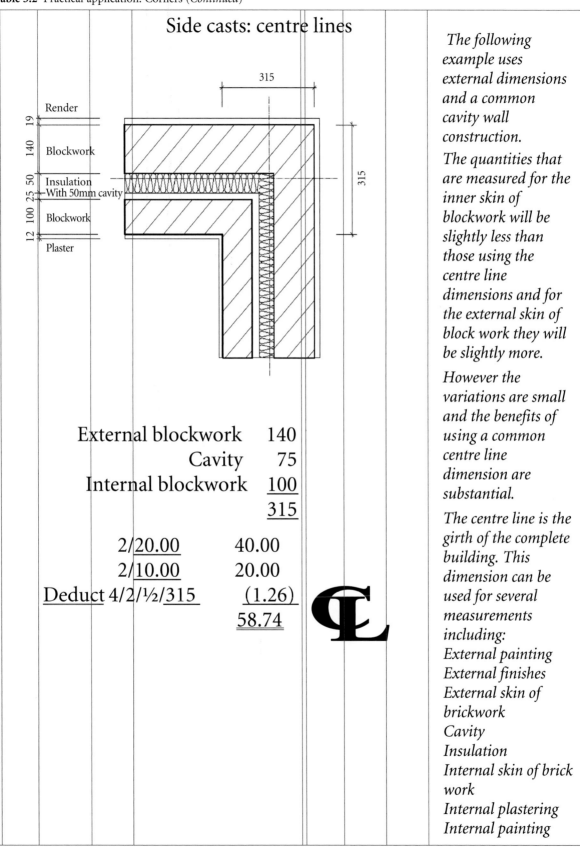

Side casts: centre lines

315

315

Render

19
140
Blockwork

50
25
Insulation
With 50mm cavity

100
Blockwork

12
Plaster

External blockwork	140
Cavity	75
Internal blockwork	100
	315

2/20.00	40.00
2/10.00	20.00
Deduct 4/2/½/315	(1.26)
	58.74

ℂ𝕃

The following example uses external dimensions and a common cavity wall construction.

The quantities that are measured for the inner skin of blockwork will be slightly less than those using the centre line dimensions and for the external skin of block work they will be slightly more.

However the variations are small and the benefits of using a common centre line dimension are substantial.

The centre line is the girth of the complete building. This dimension can be used for several measurements including:
External painting
External finishes
External skin of brickwork
Cavity
Insulation
Internal skin of brick work
Internal plastering
Internal painting

Table 3.2 Practical application: Corners (*Continued*)

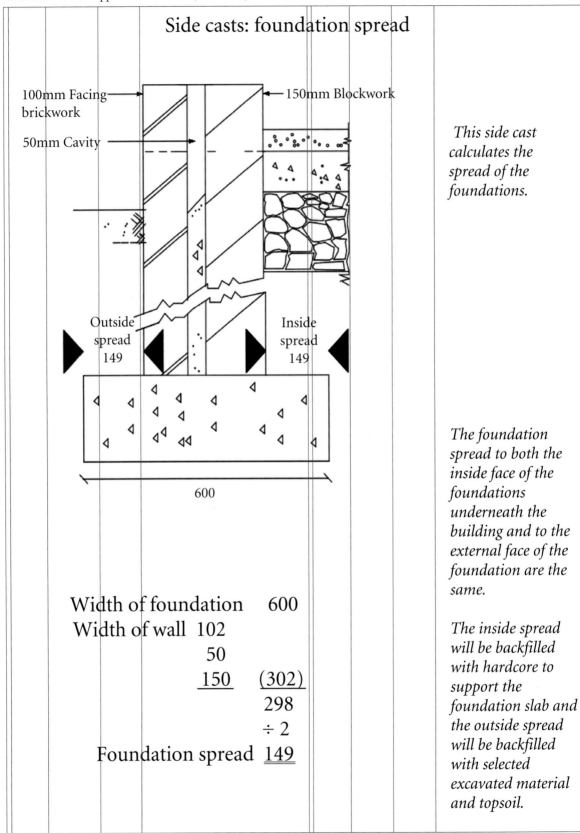

Side casts: foundation spread

100mm Facing brickwork

50mm Cavity

150mm Blockwork

Outside spread 149

Inside spread 149

600

This side cast calculates the spread of the foundations.

The foundation spread to both the inside face of the foundations underneath the building and to the external face of the foundation are the same.

The inside spread will be backfilled with hardcore to support the foundation slab and the outside spread will be backfilled with selected excavated material and topsoil.

Width of foundation   600
Width of wall 102
                50
               150   (302)
                      298
                      ÷ 2
Foundation spread 149

3.4   SELF-ASSESSMENT EXERCISE: INTERNAL AND EXTERNAL DIMENSIONS

Measure the centre line using the plan on Drawing SDCO/2/E3/1 in Appendix 3. Please prepare your own measurement using blank double dimension paper in Appendix 1a and also prepare a query sheet of problems that you have encountered. Compare your own work with the proposed solution included in Appendix 3 (Table E3.1) and self-assess your work on the assessment sheet in Appendix 1b.

To provide further assistance there are dedicated websites at http://ostrowski quantities.com and at Wiley Blackwell (http://www.wiley.com/go/ostrowski/ measurement). It is hoped that the provision of this will go some way towards explaining the concepts and principles more clearly than using the printed word alone.

4 Substructure

4.1 Measurement information
- Drawings
- Specification
- Query sheet

4.2 Technology

4.3 Practical application: Substructure to doctors' surgery

4.4 Self-assessment exercise: Trench foundations

4.1 MEASUREMENT INFORMATION

Drawings

See Drawings SDCO/2/4/1 (substructure plan) and SDCO/2/4/2 (substructure section).

Measurement Using the New Rules of Measurement, First Edition. Sean D.C. Ostrowski.
© 2013 John Wiley & Sons, Ltd. Published 2013 by John Wiley & Sons, Ltd.

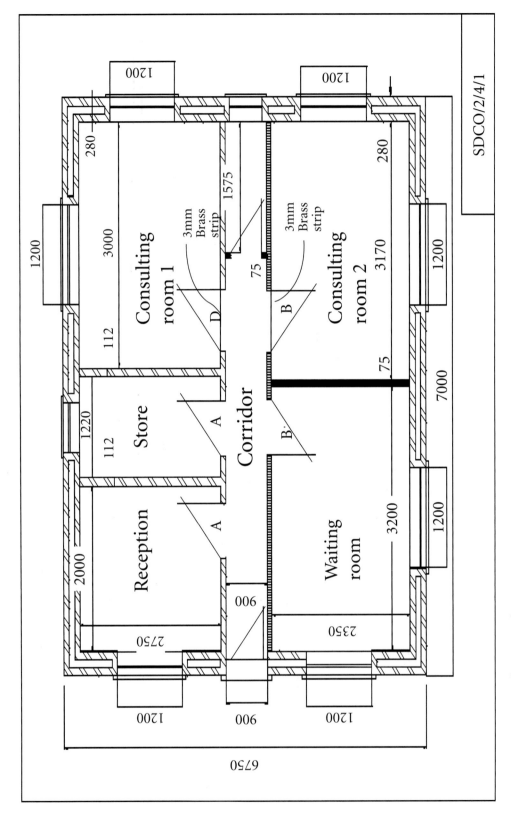

Drawing SDCO/2/4/1 Substructure plan.

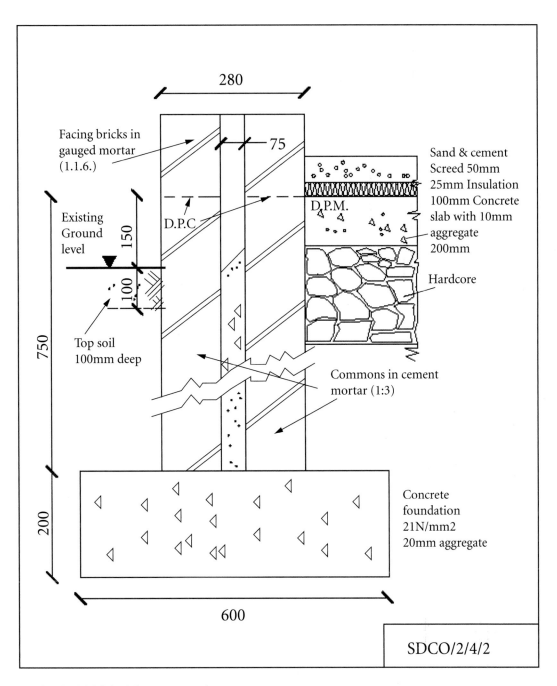

280

Facing bricks in
gauged mortar
(1.1.6.)

75

Sand & cement
Screed 50mm
25mm Insulation
100mm Concrete
slab with 10mm
aggregate
200mm

D.P.M.

Existing
Ground
level

150

D.P.C

100

Top soil
100mm deep

Hardcore

750

Commons in cement
mortar (1:3)

Concrete
foundation
21N/mm2
20mm aggregate

200

600

SDCO/2/4/2

Drawing SDCO/2/4/2 Substructure section.

Specification

Table 4.1 Substructure specification.

Specification S4: Doctors' surgery
All surplus soil to be stored on site 50m from area of excavation.
All surplus excavated material to be removed from site.
Hardcore to be broken brick or stone blinded with 50mm sand.
Plain concrete in-situ foundations to be ready mixed concrete, BS 5328:2 Section 4, designated mix 20/25, 25mm aggregate.
Reinforced concrete in-situ bed to be to be ready mixed concrete, BS 5328:2 Section 4, designated mix 20/25, 20mm aggregate.
Bed reinforced with two layers of BS 4483, A142, 2.22kg/m², mesh.
Brickwork below ground to be common bricks, BS 3921, stretcher bond, with cement mortar (1:3).
Brickwork above ground to be facing bricks, BS 3921, Prime Cost (PC) £500/1000, stretcher bond, with gauged mortar (1:1:6).
Cavity to be 80mm wide formed with polypropylene ties, 3 nr per m².
Cavity below DPC to be filled with weak mix (1:10) concrete.
Damp proof membrane to be 1000 gauge polythene with 100mm lapped joints.
Damp proof course, Hyload, proprietary, or equal and approved, pitch polymer, BS 743, 150mm laps, bedded in cement mortar (1:3).
Cavity tray, Cavity Trays Ltd., or equal and approved Type X, polypropylene abutment tray, complete with flashings.

Query sheet

Table 4.2 Substructure query sheet.

QUERY SHEET	
QUERY **(From the QS)**	**ANSWER** **(From the Architect/Engineer)** **(Assumptions are to be confirmed by QS)**
1. Slab datum level.	Not required. Measure as section. Confirmed xx.xx.xx
2. Existing ground datum level	Not required. Confirmed xx.xx.xx
2. Scale of Section	Use figured dimensions. Confirmed xx.xx.xx
3. Is sand blinding required to hardcore fill?	Yes. Confirmed xx.x.xx
4. Backfill below hardcore to be Type 1 fill	Yes. Confirmed xx.xx.xx
5. External backfill to be excavated material	Yes. Confirmed xx.xx.xx
6. Concrete bed reinforcement	Two layers A142. Confirmed xx.xx.xx
7. Cavity tray not shown on drawings	Assumed. Confirmed xx.xx.xx
etc.	

4.2 TECHNOLOGY

One of the problems of measurement is not being able to understand or visualise how the work is built. To help to overcome this each chapter will have a short construction technology description to provide the vocabulary and explanations of the method of construction.

The substructure is defined as the work below the waterproofing works, and the superstructure is the work above the waterproofing. This can be illustrated by referring to Drawing SDCO/2/4/2. In this example the Damp Proof Membrane (DPM) between the concrete slab and the screed marks the boundary between the substructure below and the superstructure above. The concrete slab is in the substructure and the insulation and screed are in the superstructure. It is not necessary to include the measurement of the insulation or screed in the substructure. In the brickwork the boundary is at the Damp Proof Course (DPC). The external boundary is the outside face of the outer skin of brickwork and the external face of the concrete foundation.

For the substructure a trench is excavated as wide as the concrete footings or foundation. There is no need for earthwork support to the sides of the trenches (also called planking and strutting) nor for formwork because the concrete will be poured directly into the trench to form the foundations. The excavation for the slab is taken down to the formation level, the bottom, of the hardcore fill below the ground slab. The two skins of brickwork to the cavity wall are built to and include the DPC level. The cavity is tied together with cavity ties and then filled with weak mix concrete to ensure the brickwork is waterproof below ground level. Both skins of the cavity wall are built with inexpensive 'common' bricks. The outer skin of brickwork will be exposed above the top surface of the topsoil and the common brick is replaced with higher quality 'facing' brick specified by the architect. Diagrams 4.1 and 4.2 show the spread of the foundations. The area above the trench foundation on the inside face is called the 'inside spread' of the foundation and the area above the trench foundation on the outside face is called the 'outside spread'.

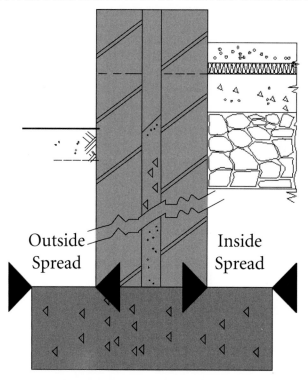

Diagram 4.1 Foundation spread: section.

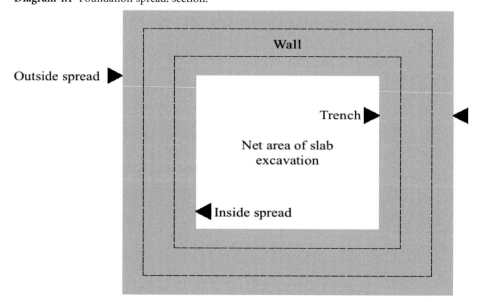

Diagram 4.2 Foundation spread: plan.

The trench is the width of the footings, which is the thickness of the wall plus the inside and outside spread. Once the trenches have been excavated, only the net area of the slab remains to be excavated. The net area of the slab is the external area of the building less the thickness of the wall and the width of the inside spread. The calculations are shown in the dimensions that are set out in the measurement that follows.

Diagram 4.3 shows the different types of backfill to the inside and outside spread of the foundations. The inside spread, excavated area (1), between the top of the foundation and the soffit of the slab, is backfilled with hardcore to support the slab. The outside spread, excavated area (2), above the foundations and below the topsoil, is backfilled with selected excavated material. It is outside the substructure area and is part of the external works. However, it is usually included in the substructure as all the information and dimensions are available. The query sheet should make plain that this is included in the substructure and is not duplicated when the external works are measured at some later stage.

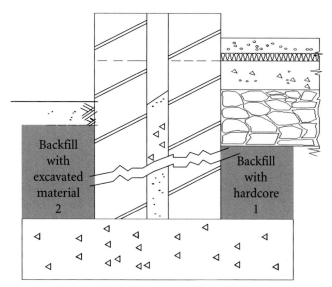

Diagram 4.3 Backfill to trenches.

The hardcore is laid and compacted and blinded with sand to receive the concrete slab. The damp proof membrane is laid and fixed on top of the concrete slab.

Table 4.3 Practical application: Substructure to doctors' surgery.

			Substructure SUBSTRUCTURE DOCTORS' SURGERY Drawings SDCO/2/4/1 Plan SDCO/2/4/2 Section Specification S1 *The title page includes the trade name, the name of the contract, the full drawing schedule with dated revisions, the dated specification, the name of the measurer and the date.*				*The first thing to do is a series of preliminary calculations or 'side casts'. They provide basic information like the centre line, length and depth of the trench and the depth of excavation for the slab and the 'spread' of the foundations.* *These calculations are shown on the double dimension paper as follows and are accompanied by explanations of the calculations.* *These explanations are to assist in the understanding of the measurement process. Such explanations would not normally appear in the taking-off of the dimensions. The only descriptions that are necessary are the NRM descriptions.*

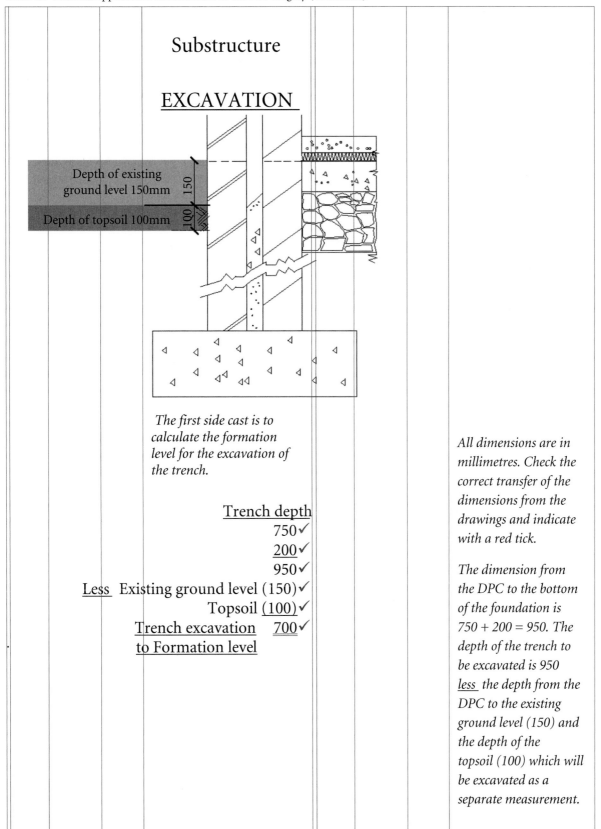

Substructure

EXCAVATION

Depth of existing ground level 150mm

Depth of topsoil 100mm

150

100

The first side cast is to calculate the formation level for the excavation of the trench.

Trench depth
750 ✓
200 ✓
950 ✓
Less Existing ground level (150) ✓
Topsoil (100) ✓
Trench excavation   700 ✓
to Formation level

All dimensions are in millimetres. Check the correct transfer of the dimensions from the drawings and indicate with a red tick.

The dimension from the DPC to the bottom of the foundation is 750 + 200 = 950. The depth of the trench to be excavated is 950 *less* the depth from the DPC to the existing ground level (150) and the depth of the topsoil (100) which will be excavated as a separate measurement.

Substructure

EXCAVATION

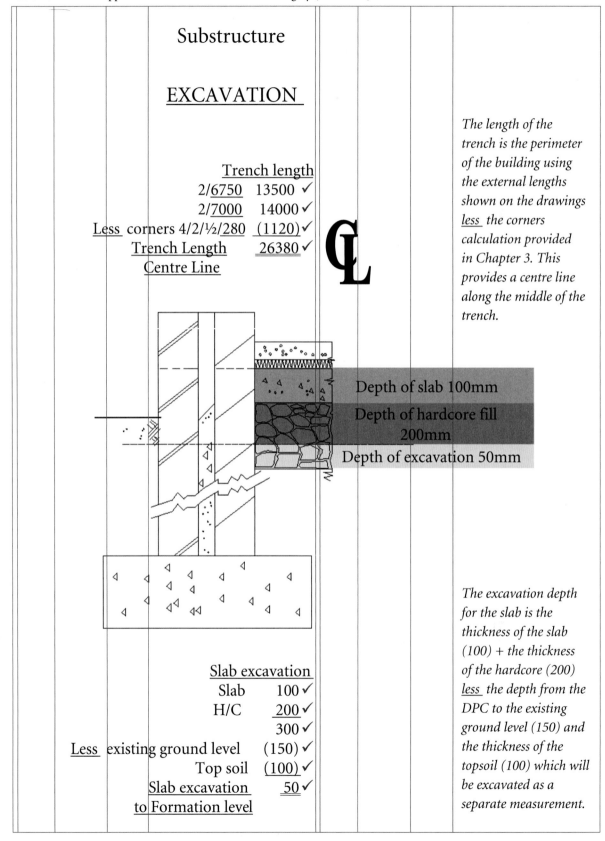

The length of the trench is the perimeter of the building using the external lengths shown on the drawings *less* the corners calculation provided in Chapter 3. This provides a centre line along the middle of the trench.

Trench length

2/6750	13500	✓
2/7000	14000	✓
Less corners 4/2/½/280	(1120)	✓
Trench Length	26380	✓
Centre Line		

C L

Depth of slab 100mm

Depth of hardcore fill 200mm

Depth of excavation 50mm

The excavation depth for the slab is the thickness of the slab (100) + the thickness of the hardcore (200) *less* the depth from the DPC to the existing ground level (150) and the thickness of the topsoil (100) which will be excavated as a separate measurement.

Slab excavation

Slab	100	✓
H/C	200	✓
	300	✓
Less existing ground level	(150)	✓
Top soil	(100)	✓
Slab excavation	50	✓
to Formation level		

Table 4.3 Practical application: Substructure to doctors' surgery (*Continued*)

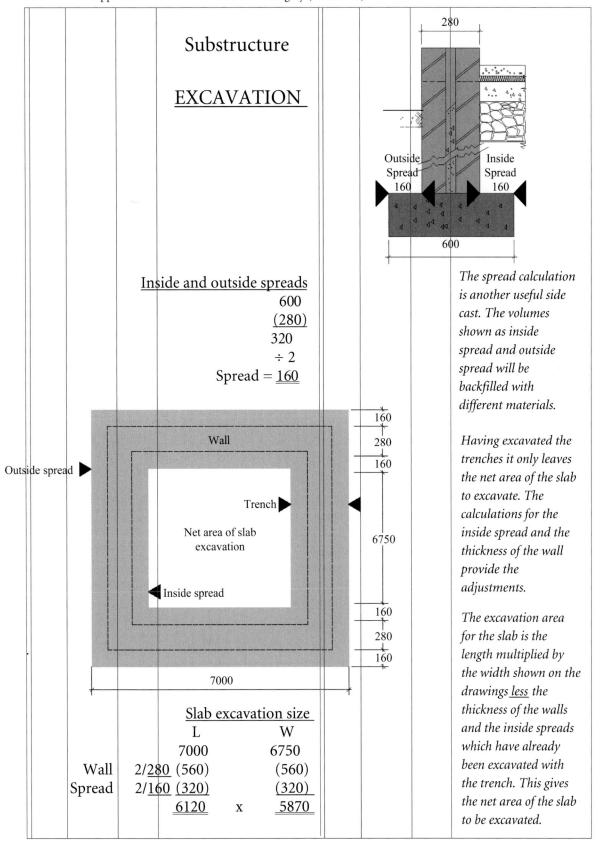

Substructure

EXCAVATION

280

Outside Spread 160   Inside Spread 160

600

Inside and outside spreads

600
(280)
320
÷ 2
Spread = 160

160
280
160

Wall

Outside spread ▶

Trench ▶   ◀

Net area of slab excavation

6750

◀ Inside spread

160
280
160

7000

Slab excavation size

		L	W
		7000	6750
Wall	2/280	(560)	(560)
Spread	2/160	(320)	(320)
		6120	5870

6120   x   5870

The spread calculation is another useful side cast. The volumes shown as inside spread and outside spread will be backfilled with different materials.

Having excavated the trenches it only leaves the net area of the slab to excavate. The calculations for the inside spread and the thickness of the wall provide the adjustments.

The excavation area for the slab is the length multiplied by the width shown on the drawings less the thickness of the walls and the inside spreads which have already been excavated with the trench. This gives the net area of the slab to be excavated.

Table 4.3 Practical application: Substructure to doctors' surgery (*Continued*)

			Substructure				
			EXCAVATION				
Item			Preliminary siteworks [5.1				We can now commence the measurement, starting with the relevant items at the beginning of Section 5 of the NRM 2.
2/ 1			Trial pits (Say 2 nr) [5.1.2				Where no information is available an assumption is made. This should be put on a query sheet for confirmation or alteration by the structural engineer.

The thick black line brackets items together. Here there is only one description and one dimension per item. Later there will be several descriptions and/or dimensions bracketed together.

NRM 2 reference numbers are included with each description in order to illustrate the process. The references are not necessary because the measurement should be in accordance with the NRM 2. |

Table 4.3 Practical application: Substructure to doctors' surgery (*Continued*)

<u>EXCAVATION</u>

		L		W
		7000		6750
Outside spread	2/160	320		320
		7320	x	7070

7.32		Site clearance.
7.07	51.75	[5.4.1

&

Remove topsoil, ave. 100mm depth, stored on site

[5.5.2.1

&

Retain topsoil on site, to temporary spoil heap, 50m distance

[5.10.1.1

51.75 x 0.10 = 5.18 m³

The area for the removal of the undergrowth is the length and width of the external dimensions shown on the drawings plus the outside spread of the foundations.

The dimensions for inclusion in the 'dimension' column are measured in accordance with NRM 2 which is to two decimal points. The dimensions have also been 'squared' to show the area in square metres and the result is included in the squaring column.

The two items of site clearance and removal of topsoil are measured as superficial areas and have common dimensions.

Retaining the topsoil on site is measured as cubic metres. The superficial area is multiplied by the depth to make it cubic metres as NRM 2.

At this stage we can check the specification to see that we have correctly measured what is required. The specification has been highlighted to show that only the first item has been measured to date.

The drawings should also be highlighted as each item of measurement is completed. These two items, the 'marking-up' of the specification and the drawings, are a very effective management and checking tool to ensure that all the work has been measured.

Specification S4 **Doctors' surgery**
All surplus soil to be stored on site 50m from area of excavation.
All surplus excavated material to be removed from site.
Hardcore to be broken brick or stone blinded with 50mm sand.
Plain concrete in-situ foundations to be ready mixed concrete, BS 5328:2 Section 4, designated mix 20/25, 20N/mm^2, 25mm aggregate.
Reinforced concrete in-situ bed to be to be ready mixed concrete, BS 5328:2 Section 4, designated mix 20/25, 20N/mm^2, 20mm aggregate.
Bed reinforced with two layers of BS 4483, A142, 2.22kg/m^2, mesh.
Brickwork below ground to be common bricks, BS 3921, stretcher bond, with cement mortar (1:3).
Brickwork above ground to be facing bricks, BS 3921, Prime Cost (PC) £500/1000, stretcher bond, with gauged mortar (1:1:6).
Cavity to be 75mm wide formed with polypropylene ties, 3 nr per m^2.
Cavity below DPC to be filled with weak mix (1:10) concrete.
Damp proof membrane to be 1000 gauge polythene with 100mm lapped joints.
Damp proof course, Hyload, proprietary, or equal and approved, pitch polymer, BS 743, 150mm laps, bedded in cement mortar (1:3).
Cavity tray, Cavity Trays Ltd., or equal and approved Type X, polypropylene abutment tray, complete with flashings.

Table 4.3 Practical application: Substructure to doctors' surgery (*Continued*)

Substructure

Dimensions	Description	Commentary
26.38 .60 .70	Excavation, foundation excavation, ne 2m, trenches [5.6.2.1	*The excavation for the trench is measured in m^3 and is in the order length x width x depth. The dimensions come from the side casts calculated earlier.*
Item	Disposal of excavated material off site [5.9.1.2	
Item	Disposal of ground water, depth below original gwl ------- [5.9.1.1	*Where ground water exists the removal of it is included as an 'Item'. The depth below the existing ground water level is to be stated where available.*
26.38 .60 .20	Plain in-situ concrete, mix 20/25, 20mm aggregate, horizontal work, < 300 thick, in structures. [11.2.1.2	*This is the footing to the trench. The usefulness of the centre line calculation can now be seen as it is used for several different measurements.*
2/26.38 .75	Wall, ½b thick, in commons, cement mortar (1:3), skins of hollow walls, stretcher bond [14.1.1.1	*There are two skins of brickwork in the cavity wall so the measurement is doubled.*

Say 3 courses of bricks 3/75 = <u>225</u>

Dimensions	Description	Commentary

26.38
.23
	<u>Ddt</u>
	Wall, ½b th, in commons, c/m (1:3), stretcher bond
	                    [14.1.1.1
	
	&
	
	<u>Add</u>
	Wall, ½b th, in facings (PC £500/1000), gauged mortar (1:1:6), abd, stretcher bond
	                    [14.1.1.1
		Replace say three courses of common bricks with facing bricks as an allowance of 150 for undulations in the topsoil.
		
		The brackets indicate that both descriptions have common dimensions.

<table>
<tr><td colspan="3" align="center">Substructure</td><td></td><td></td></tr>
</table>

			280			The width of the cavity is the thickness of the wall *less* both skins of brickwork.
	Less	2/102.50 mm (205)				
		Width of cavity 75				

| 26.38 | | | Forming cavity, 75mm wide, formed with S/S butterfly ties, BS 1243, 5/m² | | | |
| .70 | 18.47 | | [14.14.1.1 | | | |

&

		Plain in-situ concrete, (1:10), mass concrete, filling cavities			Concrete is measured in m³ so the superficial dimensions in m² are converted to m³.
		[11.1.1.3			
		18.47 x 0.08 = 1.48m³			

| 2/26.38 | | DPC, < 300mm wide, pitch polymer, BS 743, bedded in c/m (1:3), horizontal | | | |
| | | [14.16.1.3 | | | |

| 26.38 | | Pre-formed proprietary cavity tray, bedded in c/m (1:3), horizontal | | | Not shown on drawing but confirmed in query sheet. |
| | | [14.18.1.3 | | | |

Frequent checks and highlighting of the specification ensure that all the work has been measured.

Specification S4 **Doctors' surgery**
All surplus soil to be stored on site 50m from area of excavation.
All surplus excavated material to be removed from site.
Hardcore to be broken brick or stone blinded with 50mm sand.
Plain concrete in-situ foundations to be ready mixed concrete, BS 5328:2 Section 4, designated mix 20/25, 20N/mm^2, 25mm aggregate.
Reinforced concrete in-situ bed to be to be ready mixed concrete, BS 5328:2 Section 4, designated mix 20/25, 20N/mm^2, 20mm aggregate.
Bed reinforced with two layers of BS 4483, A142, 2.22kg/m^2, mesh.
Brickwork below ground to be common bricks, BS 3921, stretcher bond, with cement mortar (1:3).
Brickwork above ground to be facing bricks, BS 3921, Prime Cost (PC) £500/1000, stretcher bond, with gauged mortar (1:1:6).
Cavity to be 75mm wide formed with polypropylene ties, 3 nr per m^2.
Cavity below DPC to be filled with weak mix (1:10) concrete.
Damp proof membrane to be 1000 gauge polythene with 100mm lapped joints.
Damp proof course, Hyload, proprietary, or equal and approved, pitch polymer, BS 743, 150mm laps, bedded in cement mortar (1:3).
Cavity tray, Cavity Trays Ltd., or equal and approved Type X, polypropylene abutment tray, complete with flashings.

Substructure

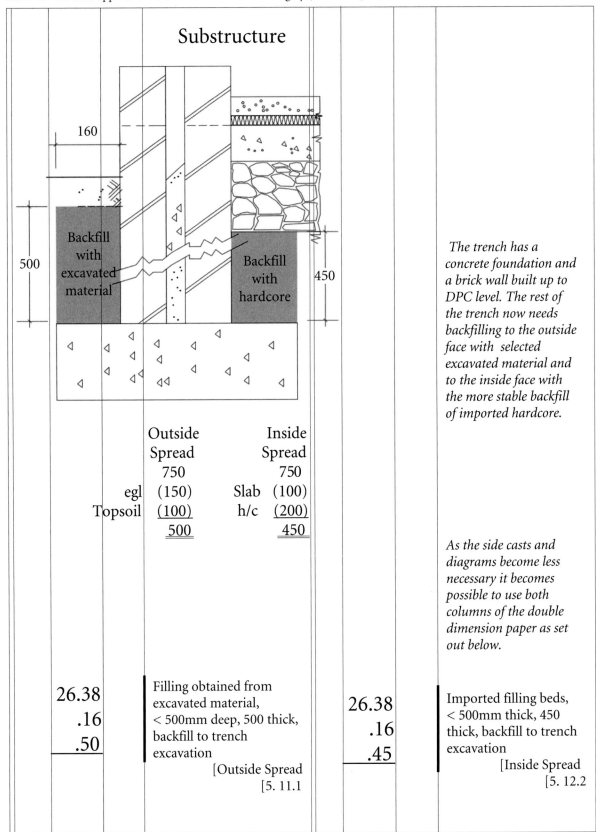

160	
500	Backfill with excavated material
	Backfill with hardcore
	450

The trench has a concrete foundation and a brick wall built up to DPC level. The rest of the trench now needs backfilling to the outside face with selected excavated material and to the inside face with the more stable backfill of imported hardcore.

	Outside Spread		Inside Spread
	750		750
egl	(150)	Slab	(100)
Topsoil	(100)	h/c	(200)
	500		450

As the side casts and diagrams become less necessary it becomes possible to use both columns of the double dimension paper as set out below.

26.38	Filling obtained from	26.38	Imported filling beds,
.16	excavated material,	.16	< 500mm thick, 450
.50	< 500mm deep, 500 thick,	.45	thick, backfill to trench
	backfill to trench		excavation
	excavation		[Inside Spread
	[Outside Spread		[5. 12.2
	[5. 11.1		

Substructure

EXCAVATION

Slab excavation
Slab		100
H/C		200
		300
Less	existing ground level	(150)
	Top soil	(100)
Formation Level Depth		50

Slab excavation size

	L	W
	7000	6750
Wall	2/280 (560)	(560)
Spread	2/160 (320)	(320)
	6120 x	5870

6.12	Bulk excavation, < 2m
5.87	deep, reduced levels
.05	5.6.1.1

The slab excavation side casts are repeated here for convenience.

Depth of slab 100mm

Depth of hardcore fill 200mm

Depth of excavation 50mm

Table 4.3 Practical application: Substructure to doctors' surgery (*Continued*)

Substructure

Slab size

		L	W	*The area of the slab is the net area inside the brick wall, so the thickness of the wall is deducted from the dimensions for the length and the width of the building.*
		7000	6750	
Wall	2/280	(560)	(560)	
		6440 x	6190	

6.44
6.19 39.86

Damp proof membrane, 1000g, polythene, horizontal

[5.16.2.1

Upstands for membranes where they tuck into the side of the slab are included.

&

Mesh, BS 4482, A142, 2.22km/m², minimum 100mm laps

[11.37.1

The laps are assumed and should be confirmed on a query sheet.

39.86 x 2 = 79.72 m²

6.44
6.19
.05

Imported filling, blinding beds, sand, <50mm thick, 50mm thick

[5.12.1

6.44
6.19
.20

Imported filling, < 500mm thick, hardcore, in 150mm layers, fill to excavation

[5.12.2.1

6.44
6.19
.10

Reinforced ready mixed concrete, designated mix C20/25, 20mm aggregate, horizontal work, < 300mm thick, in structures

[11.2.1.2

The specification for the concrete should be confirmed on the query sheet.

Both columns of the double dimension paper should be used. In this example the right hand column has often been used for additional notes, diagrams and clarification in several areas.

End of Substructure

Check and highlight the specification at the end to ensure that all the work has been measured. In this example the insulation below the screed but above the damp proof membrane (DPM) will be measured in the superstructure in a later section.

Specification S4 **Doctors' surgery**
All surplus soil to be stored on site 50m from area of excavation.
All surplus excavated material to be removed from site.
Hardcore to be broken brick or stone blinded with 50mm sand.
Plain concrete in-situ foundations to be ready mixed concrete, BS 5328:2 Section 4, designated mix 20/25, 20N/mm², 25mm aggregate.
Reinforced concrete in-situ bed to be to be ready mixed concrete, BS 5328:2 Section 4, designated mix 20/25, 20N/mm², 20mm aggregate.
Bed reinforced with two layers of BS 4483, A142, 2.22kg/m², mesh.
Brickwork below ground to be common bricks, BS 3921, stretcher bond, with cement mortar (1:3).
Brickwork above ground to be facing bricks, BS 3921, Prime Cost (PC) £500/1000, stretcher bond, with gauged mortar (1:1:6).
Cavity to be 75mm wide formed with polypropylene ties, 3 nr per m².
Cavity below DPC to be filled with weak mix (1:10) concrete.
Damp proof membrane to be 1000 gauge polythene with 100mm lapped joints.
Damp proof course, Hyload, proprietary, or equal and approved, pitch polymer, BS 743, 150mm laps, bedded in cement mortar (1:3).
Cavity tray, Cavity Trays Ltd., or equal and approved Type X, polypropylene abutment tray, complete with flashings.

4.4   SELF-ASSESSMENT EXERCISE: TRENCH FOUNDATIONS

Using the example shown in this chapter as a guide, measure the trench excavation using the plan and section on Drawing SDCO/2/E4/1 in Appendix 4. Please prepare your own measurement using blank double dimension paper in Appendix 1a and also prepare a query sheet of problems that you have encountered. Compare your own work with the proposed solution included in Appendix 4 (Table E4.1) and self-assess your work on the assessment sheet in Appendix 1b.

To provide further assistance there are dedicated websites at http://ostrowski quantities.com and at Wiley Blackwell (http://www.wiley.com/go/ostrowski/ measurement). It is hoped that the provision of this will go some way towards explaining the concepts and principles more clearly than using the printed word alone.

5 Basement

Measurement Using the New Rules of Measurement, First Edition. Sean D.C. Ostrowski.
© 2013 John Wiley & Sons, Ltd. Published 2013 by John Wiley & Sons, Ltd.

5.1   MEASUREMENT INFORMATION

Drawings

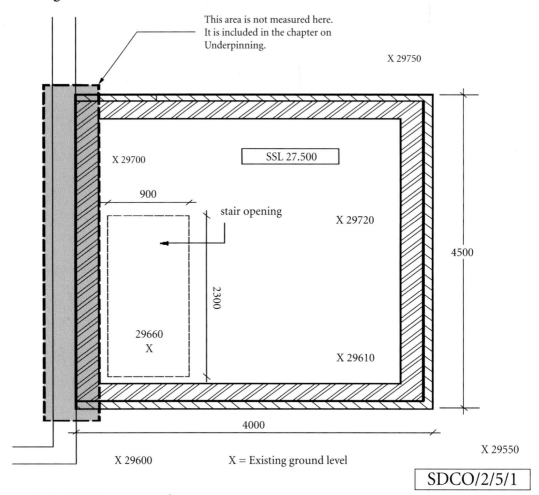

Drawing SDCO/2/5/1 Basement plan.

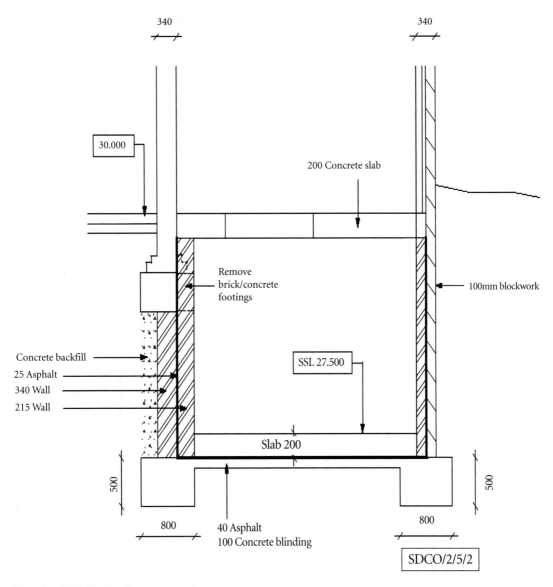

340

340

30.000

200 Concrete slab

Remove
brick/concrete
footings

100mm blockwork

Concrete backfill

25 Asphalt

340 Wall

215 Wall

SSL 27.500

Slab 200

500

500

800

800

40 Asphalt
100 Concrete blinding

SDCO/2/5/2

Drawing SDCO/2/5/2 Basement section.

Specification

Table 5.1 Basement: sloping site specification.

Specification S5: Basement
200mm of vegetable soil to be stripped, and filled to planters, 10m from the building.
Precontract water level determined at 28.500.
All surplus excavated material to be removed from site.
Hardcore to be broken brick or stone blinded with 50mm concrete.
Mass concrete foundations and blinding beds to BS12, designed mix, grade C25, sulphate resisting cement, 40mm aggregate.
Loading slab laid over asphalt to be ready mixed concrete, BS 5328, designated mix 25/30, $25N/mm^2$, 30mm aggregate.
Reinforcement details to follow. Allow 1.5% by volume of concrete.
Basement walls to be Class B engineering bricks, SHP Warnham bricks in cement mortar (1:4), 215mm thick.
Protective external cladding to be 100mm dense concrete blocks ($7N/mm^2$) in cement mortar (1:3), built against asphalt tanking.
Asphalt tanking to existing and new brick walls to be mastic asphalt to BS 6925, type T 1097, 20mm thick, two coat work to brickwork base, subsequently covered.
Asphalt tanking to concrete bed to be 40mm two coat asphalt as before described.
Serviseal external PVC waterstop, centrebulb type, heat welded joints, cast into concrete, Serviseal type 195. Plus angles and intersections.

Query sheet

Table 5.2 Basement query sheet.

QUERY (From the QS)	ANSWER (From the Architect/Engineer) (Assumptions are to be confirmed by QS)
1.Spoil to be removed from site	Confirmed xx.xx.11
2. Underpinning	Excluded. Measure elsewhere. Confirmed xx.xx.xx
3. Scale of Section	Use figured dimensions. Confirmed xx.xx.xx
4. Working space	Assumed. Confirmed xx.x.xx
5. Backfill below hardcore to be Type 1 fill	Yes. Confirmed xx.xx.xx
6. Backfill to be mass concrete.	Yes. Confirmed xx.xx.xx
7. Concrete bed reinforcement	Assumed 1.5%. Confirmed xx.xx.xx
8. Specification for Warnham bricks	PC £500/1000
9. Assume stretcher bond	Confirmed xx.xx.xx
10. DPC specification	Assumed. Confirmed xx.xx.xx
11. Proprietary tanking	Assumed. Confirmed xx.xx.xx
12. Waterbar not used as tanking used	Confirmed xx.xx.xx
13. Reinforcement	Provisional allowances confirmed

5.2   TECHNOLOGY

This basement is being excavated immediately adjacent to an existing building. This requires underpinning to the adjacent building before the construction work to the basement is undertaken. The underpinning is described in Chapter 7.

The existing precontract water level is below the datum of the slab so allowances will need to be included for ground water in addition to surface water.

The brickwork will be built in restricted spaces and adjacent to the excavation. The descriptions will reflect these restrictions.

The asphalt will be applied to the new brickwork walls and the preparation of the walls, by raking out the pointing, will be included in the descriptions.

Table 5.3 Practical application: Brick basement to adjacent building.

			Substructure				
			BASEMENT				
			EXISTING BUILDING				
			Drawings				
			SDCO/2/5/1 Plan				
			SDCO/2/5/2 Section				
			Specification S5				
							The title page includes the trade name, the name of the contract, the full drawing schedule with dated revisions, the dated specification, the name of the measurer and the date.
			SDCO				*This section is also used for the exercise for BQ preparation in Chapter 24 and therefore the quantities are squared.*
			May 2011				

Table 5.3 Practical application: Brick basement to adjacent building (*Continued*)

Basement

EXCAVATION

Average existing gl

29700	✓
29660	✓
29600	✓
29720	✓
29610	✓
29750	✓
29550	✓
207590	✓
÷ 7	✓

Ave egl = 29656 ✓

	29656	✓	
SSL	27500	✓	
Slab	200 ✓		
Asphalt	40 ✓		
Blinding	100 ✓	(340) ✓	
Topsoil	200 ✓	(27360) ✓	
EXCAVATION	2296 ✓		

Extra over for excavation below gwl

Precontract gwl	28500	✓
Formation level	(27360)	✓
Excavation below gwl	1140	✓

The first side cast is to calculate the existing ground level from the existing datum levels provided on the drawing.

All the existing levels have been used although some of them lie outside the curtledge of the building.

All dimensions are in millimetres. Check the correct transfer of the dimensions from the drawings and indicate with a red tick.

egl: existing ground level

gwl: ground water level

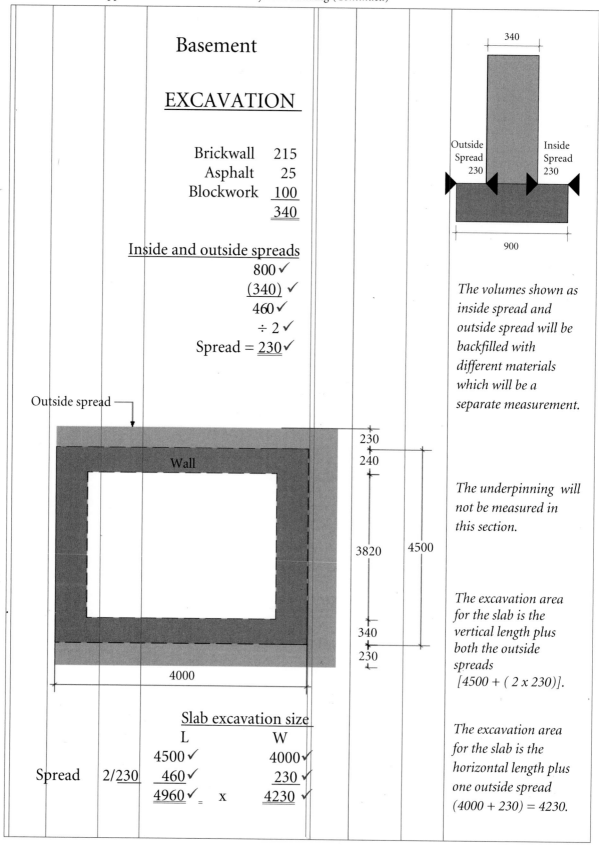

Basement

EXCAVATION

Brickwall	215
Asphalt	25
Blockwork	100
	340

Inside and outside spreads

800 ✓
(340) ✓
460 ✓
÷ 2 ✓
Spread = 230 ✓

Outside spread

Wall

230
240

3820   4500

340
230

4000

Slab excavation size

	L		W
	4500 ✓		4000 ✓
Spread	2/230 460 ✓		230 ✓
	4960 ✓ =	x	4230 ✓

340

Outside Spread 230   Inside Spread 230

900

The volumes shown as inside spread and outside spread will be backfilled with different materials which will be a separate measurement.

The underpinning will not be measured in this section.

The excavation area for the slab is the vertical length plus both the outside spreads [4500 + (2 x 230)].

The excavation area for the slab is the horizontal length plus one outside spread (4000 + 230) = 4230.

Table 5.3 Practical application: Brick basement to adjacent building (*Continued*)

			Substructure				
			EXCAVATION				
Item	Item		Preliminary siteworks [5.1				NRM reference numbers are included with each description only in order to illustrate the process. The references are not necessary because the measurement should be in accordance with the NRM 2.
2/ 1	2		Trial pits (Say 2 nr) [5.1.2				
4.23 ✓ 4.96 ✓	20.98 ✓		Site clearance [5.4.1 & Remove topsoil, ave. 200mm, stored on site [5.5.2.1 & Filling, topsoil, from excavated material, < 500mm, from spoil heaps, spread & levelled on site 10m from the building to planter beds & the like [5.11.1.1.1.3 $20.98 \times 0.20 = \underline{4.12\text{m}^3}$				The two items site clearance and removal of topsoil are measured as superficial areas and have common dimensions. Filling with the topsoil to the planter beds on site is measured as cubic metres. The superficial area is multiplied by the depth to make it cubic metres.

Table 5.3 Practical application: Brick basement to adjacent building (*Continued*)

Substructure

4.23 ✓		Bulk excavation, basements, 2–4 m deep [5.6.1.2			The dimensions come from the side casts calculated earlier.
4.96 ✓					
2.30 ✓					
	48.25 ✓				The Item for disposal of material includes all the excavation, not just bulk excavation.
Item		Disposal, excavated material off site [5.9.2			
	Item				
4.23 ✓		Extra over all types of excavation, irrespective of depth, excavating below gwl [5.7.1.3			Where ground water exists the removal of it is included as an 'Item'. The depth below the existing ground water level is to be stated where available.
4.96 ✓					
1.14 ✓					
	23.92 ✓				
Item		Disposal of ground water, 1140 below egwl [5.9.1.1			
	Item				This is the footing for the trench using the centre line calculation .

Trenches

2/4000 ✓	8000 ✓
	4500 ✓
	12500 ✓
Less 2/2/½/340 ✓	(680) ✓
	11820 ✓

Because the underpinning is not included in this section there are only three sides to measure and the centre line adjustment is for two corners only.

Depth

	500 ✓
Less blinding	(100) ✓
Depth	400 ✓

The formation level to the bottom of the excavation for the slab includes the blinding, which is then deducted from the depth of the trench for the footings.

11.82 ✓		Foundation excavation, trenches, <2.00m deep [5.6.2.1			
.80 ✓					
.40 ✓					
	3.78 ✓	&			Working space is not a measureable item in the NRM.
		EO excavation below gwl [5.7.1.3			

C L

Table 5.3 Practical application: Brick basement to adjacent building (*Continued*)

Substructure

4.23 ✓ 4.96 ✓ .10 ✓ 2.10	In-situ concrete, grade C10, cement to BS 12 sulphate resisting, 40 mm aggregate, horizontal work, < 300 thick, in blinding [11.2.1.1			*The concrete to the blinding bed, the footings and the ground slab is the same specification.*
11.82 ✓ .80 ✓ .40 ✓ 3.78 ✓	Ditto, >300thick in structures, trenches [11.2.2.2.1			*'Ditto' repeats the description before the change in the description.*

Concrete slab

4000 ✓ x 4500 ✓
2/215(430) ✓ 2/340 (680) ✓
3470 ✓ x 3820 ✓

3.47 ✓ 3.82 ✓ .20 ✓ 2.73 ✓	Ditto, Grade C25 < 300 thick in structures, 20mm aggregate, slab [9.2.1.1			*'Do' repeats the description after the change in the description.* *eg Ditto, > 300 thick, do*
4.23 ✓ 4.96 ✓ 3.57 ✓ 20.98 ✓ 3.82 ✓ 13.64 ✓ 3.83 ✓ Ddt 4.16 ✓ 2.30 ✓ 15.93 ✓ .90 ✓ (2.07) ✓ 34.64 ✓	Surface finishes to in-situ concrete, power floating [11.9.1 [blinding [slab [suspended slab [Less S/C			*The blinding bed needs power floating to receive the asphalt.* *The side casts for the suspended slab are on the following page.*

Substructure

<u>Adjustments for backfilling to external spread</u>

	2300 ✓
	(100) ✓
	2200 ✓

11.82 ✓
.23 ✓
2.20 ✓   5.98 ✓

Imported filling, hardcore, > 500mm deep, compressed in 225mm layers

[5.12.3

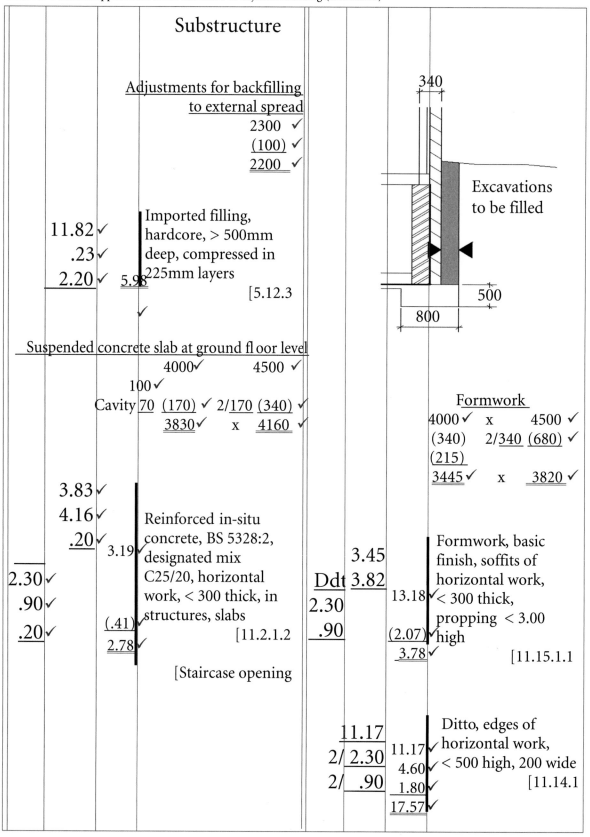

340

Excavations to be filled

500

800

<u>Suspended concrete slab at ground floor level</u>

4000 ✓   4500 ✓
100 ✓
Cavity 70 (170) ✓ 2/170 (340) ✓
3830 ✓   x   4160 ✓

<u>Formwork</u>
4000 ✓   x   4500 ✓
(340)   2/340 (680) ✓
(215)
3445 ✓   x   3820 ✓

3.83 ✓
4.16 ✓
.20 ✓   3.19 ✓

Reinforced in-situ concrete, BS 5328:2, designated mix C25/20, horizontal work, < 300 thick, in structures, slabs

[11.2.1.2

[Staircase opening

2.30 ✓
.90 ✓
.20 ✓

(.41) ✓
2.78 ✓

3.45
Ddt 3.82
2.30
.90
(2.07) ✓
3.78 ✓

13.18 ✓

Formwork, basic finish, soffits of horizontal work, < 300 thick, propping < 3.00 high

[11.15.1.1

11.17
2/ 2.30
2/ .90

11.17 ✓
4.60 ✓
1.80 ✓
17.57 ✓

Ditto, edges of horizontal work, < 500 high, 200 wide

[11.14.1

Table 5.3 Practical application: Brick basement to adjacent building (*Continued*)

			Brickwork		15.74 ✓			Walls, 215 thick, brickwork, Class B engineering bricks, SHP Warnham bricks in c/m 1:4, built against other work, bonding to other work, keyed one side [14.1.1.*.1
			Inner skin		2.50 ✓			
		2/4500	9000 ✓			39.35 ✓		
		2/4000	8000 ✓					
			17000 ✓					
Protective skin	3/100	(300) ✓						
Asphalt	2/2/25	(100) ✓						
Inner skin 4/2/½/215	(860) ✓							
		15740 ✓						
					2.30 ✓			Walls, 100 thick, blockwork, dense aggregate concrete blocks 7N/mm², vertical, in c/m 1:3, bonding to other work [14.1.2.*.4
		Height			2.70 ✓			
	SSL	30000 ✓				33.21 ✓		
		(27500) ✓						
		2500 ✓						
	Slab	(200) ✓						
		2300 ✓						
	Basement	200 ✓						
		2500 ✓						
		Protective skin			12.30 ✓			Ddt. Ditto
		4500 ✓			.49 ✓		6.03 ✓	&
	2/4000	8000 ✓						Add Walls, brickwork, ½b thick, vertical, built against other work, overhand, facework, (PC £350/1000) in g/m 1:1:6 [14.1.1.*.1
		12500 ✓						
	2/2/½/100	(200) ✓						
		12300 ✓						
		Height						
	SSL	30000 ✓						
		(27500) ✓						
		2500 ✓						
	Basement slab	200 ✓						
		2700 ✓						
		Exposed facework below DPC						DPC,<300mm wide, 'Aluminite' aluminium cored bitumen, gas retardant, 200 laps horizontal, g/m (1:1:6) [14.16.1.3
	SSL	30000 ✓						
	eagl	(29656) ✓			12.30 ✓			
		344 ✓			12.30 ✓			
	Allow below DPC	150 ✓						
		494 ✓						

			Reinforcement
	Asphalt		**RC volumes**

Less protective walls 2/100

	4500
	(200)
	4300

Fndns 3.78m^3
Slab   2.73m^3
Susp'd slab 2.78 m^3
9.29m^3
x 2%

=

.186m^3 x 7865kg/m^3
= 1.463t

4.30 ✓		Waterproofing,
2.50 ✓		coverings >500 wide,
12.30 ✓	10.25	vertical, mastic
2.70 ✓		asphalt, BS 6925,
	33.21	Type T, specification
	43.96	J20.1(b), tanking/damp

Say Straight   .8t
Bent   .4t
Links   .2t
1.4t

proofing, 20mm two
coat work,
subsequently covered,
to brickwork
[19.1.3.4.1

Reinforcement
All Provisional

	4000	4500	
2/100	(200)	(200) ✓	
	3800 ✓	4300 ✓	

Bars, BS 4449,
Grade 500C hot
rolled, mild steel,
.800 ✓ straight, 12mm
[11.33.1.1

3.80 ✓		Ditto, horizontal
4.30 ✓		40mm thick, to
	16.34	concrete

.400 ✓ Ditto, bent
[11.33.1.2

[19.1.1.4.1

.200 ✓ Ditto, links
[11.32.1.4

Both columns of the double dimension paper should be used. In this example the right hand column has often been used for additional notes, diagrams and clarification in several areas.

End of Basement

5.4 SELF-ASSESSMENT EXERCISE: BASEMENT TO OFFICE BUILDING

Measure the basement to the office building on London Rd using the plan and section on Drawing SDCO/2/E5/1 in Appendix 5. Please prepare your own measurement using blank double dimension paper in Appendix 1a and also prepare a query sheet of problems that you have encountered. Compare your own work with the proposed solution included in Appendix 5 (Table E5.1) and self-assess your work on the assessment sheet in Appendix 1b.

To provide further assistance there are dedicated websites at http://ostrowski quantities.com and at Wiley Blackwell (http://www.wiley.com/go/ostrowski/ measurement). It is hoped that the provision of this will go some way towards explaining the concepts and principles more clearly than using the printed word alone.

6 Sloping Site

6.1 Measurement information
- Drawings
- Specification
- Query sheet
6.2 Technology
6.3 Practical application: Pair of houses
6.4 Self-assessment exercise: Grid of levels

6.1 MEASUREMENT INFORMATION

Drawings

See Drawings SDCO/2/6/1 (sloping site plan) and SDCO/2/6/2 (sloping site sections).

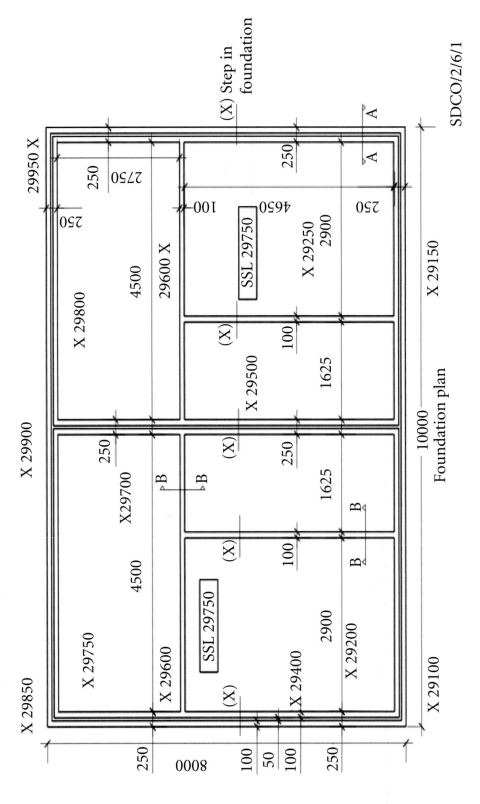

Drawing SDCO/2/6/1 Sloping site plan.

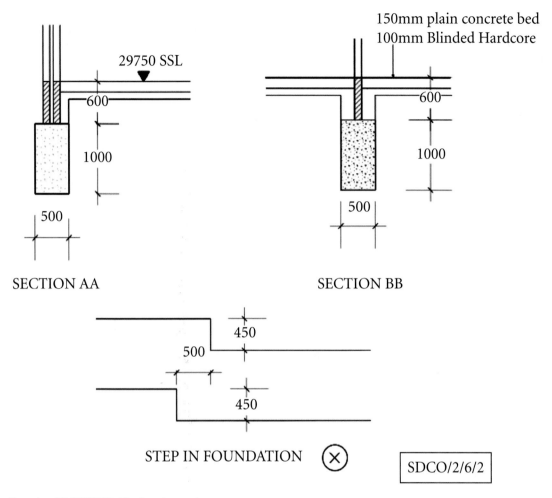

150mm plain concrete bed
100mm Blinded Hardcore

29750 SSL

600

1000

500

SECTION AA

600

1000

500

SECTION BB

450

500

450

STEP IN FOUNDATION (X)

SDCO/2/6/2

Drawing SDCO/2/6/2 Sloping site sections.

Specification

Table 6.1 Sloping site specification.

Specification S6 Sloping site
All surplus excavated material to be removed from site.
Backfill and base to be hardcore to be broken brick or stone blinded with 50mm sand.
Mass concrete foundations and blinding beds to BS12, designed mix, grade C25, sulphate resisting cement, 40mm aggregate.
Slab to be ready mixed concrete, BS 5328, 1991, section 4, designated mix 25/30, 25N/mm^2, 30mm aggregate.
Slab reinforced with two layers of BS 4483, A142, 2.22kg/m^2, mesh.
Brickwork below ground to be common bricks, BS 3921, stretcher bond, with cement mortar (1:3).
Cavity below DPC to be filled with weak mix (1:10) concrete.
Damp proof membrane to be 1000 gauge polythene with 100mm lapped joints.
Damp proof course, Hyload, proprietary, or equal and approved, pitch polymer, BS 743, 150mm laps, bedded in cement mortar (1:3).

Query sheet

Table 6.2 Sloping site query sheet.

QUERY SHEET	
SLOPING SITE	
QUERY **(From the QS)**	**ANSWER** **(From the Architect/Engineer)** **(Assumptions are to be confirmed by QS)**
1. Spoil to be removed from site	Confirmed xx.xx.xx
2. Scale of Section	Use figured dimensions. Confirmed xx.xx.xx
3. Working space	Assumed. Confirmed xx.x.xx
4. Backfill below hardcore to be Type 1 fill	Yes. Confirmed xx.xx.xx
5. Assume stretcher bond	Confirmed xx.xx.xx
6. Blockwork specification	
7. DPC specification	Assumed. Confirmed xx.xx.xx

6.2   TECHNOLOGY

The highest part of the slope is at the top of the page or the slope is from north to south. The step in the foundation is across the page, from east to west.

As indicated in Diagram 6.1, if excavation takes place to the lowest point across the whole site it will mean an excessive amount of excavation and fill. The solution is to fill the southern part of the site and excavate the northern part of the site. This will reduce both the amount of fill and the amount of excavation. The site will then be level at the formation level and the excavation for the trenches can then commence.

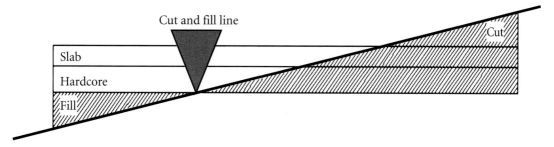

Diagram 6.1 Cut and fill line.

To determine where the cut stops and the fill starts, the cut and fill line, requires a grid of levels. This will indicate the position for the cut and fill line.

The foundations for the partitions go through the slab and adjustments will need to be made for additional trenches and backfilling

The existing precontract water level is unknown.

Table 6.3 Practical application: Pair of houses.

			Substructure				
			SLOPING SITE				
			PAIR OF BUILDINGS				
			Drawings				
			SDCO/2/6/1 Plan				
			SDCO/2/6/2 Sections & Detail				
			Specification S5				
							The title page includes the trade name, the name of the contract, the full drawing schedule with dated revisions, the dated specification, the name of the measurer and the date.
			SDCO				
			May 2012				

Table 6.3 Practical application: Pair of houses (*Continued*)

Sloping site

EXCAVATION

USING EXISTING SITE LEVELS ONLY

	Average egl	
	29750	✓
	29800	✓
	29700	✓
	29600	✓
	29600	✓
	29400	✓
	29500	✓
	29250	✓
	292000	✓
	207590	✓
	÷ 9	✓
Ave egl =	29533	✓

29533

	SSL	29750	
Slab	150		
Hardcore	100	(250)	
Formation Level	29500	(29500)	
AVE EXCAVATION		33	

This page demonstrates that simply using existing ground levels will not provide an accurate measure for a sloping site.

Using the existing datum levels provided on the drawing gives us a calculation of 33mm of excavation to reach the formation level of 29500.

However deducting 33mm from the highest existing datum 28000 gives us a level of 27967. This is still 167mm above the formation level of 29500.

Highest egl	29800
Ave excav'n	(33)
Reduced level	29667
Format'n L'vl	(29500)
Additional dig	167

There will be additional excavation to the high part of the site and fill to the low part.

Further excavation will have to take place. This demonstrates that using the overall average egl on a sloping site will not provide an accurate measure.

This calculation is not appropriate for sloping sites.

Table 6.3 Practical application: Pair of houses (*Continued*)

Sloping site

EXCAVATION

All the existing levels have been used although some of them lie outside the perimeter of the building.

Dimension checks

E–W Dimensions

2/4500 ✓	9000	2900 ✓			250 ✓
3/ 250 ✓	750 ✓	100 ✓			2900 ✓
	9750 ✓?	1625 ✓			100 ✓
		4625 ✓	2/4625 ✓	9250 ✓	1625 ✓
			3/ 250 ✓	750 ✓	250 ✓
				10000 ✓	1625 ✓
					100 ✓
					2900 ✓
					250 ✓
					10000 ✓

This indicates that the first two figured dimensions of 4500 are incorrect and should be 4625.

N–S Dimensions

250 ✓
2750 ✓
100 ✓
4650 ✓
250 ✓
8000 ✓

To establish an accurate cut level a grid of levels is required.

- Put a grid of equal sized squares over the whole site.
- This covers the whole building and can go beyond the perimeter to ensure all the building is covered.
- At the corner of each square there needs to be a datum level of the existing ground level (egl).
- This is the egl that is reasonably close to the corner.
- Where none exists extrapolate from a nearby datum level to provide the datum level at the corner.
- Plot the line of the formation level between the datum levels on the grid of levels. This is the cut and fill line.
- In the example in Diagrams 6.2 and 6.3 the datum level is 29500 and is horizontal and half way down the site.
- Split the site into the cut area and the fill area.
- Prepare a weighted schedule of levels for the cut.
  - Corner levels have a weighting of one because they are common to only one square.
  - Intermediate levels between two squares have a weighting of two because they are common to two squares.
  - Intermediate levels common to four squares have a weighting of four because they are common to four squares.
  - In this case there are 15 corners and intersections so there should be 15 datum levels.
- Add the sum of all the levels and divide by the total number of levels used to give the average egl of the cut area.
- Subtract the formation level to give the depth of cut.
- Prepare a schedule of levels for the fill.
  - Corner levels have a weighting of one because they are common to only one square.
  - Intermediate levels between two squares have a weighting of two because they are common to two squares.
  - Intermediate levels common to four squares have a weighting of four because they are common to four squares.
  - In this case there are 15 corners and intersections so there should be 15 datum levels.
- Add the sum of all the levels and divide by the total number of levels used to give the average egl of the fill area.
- Subtract the formation level to give the depth of fill. This should be a negative number to the formation level as it is higher than the average existing ground level (egl).

Single corners	1		29850
	2		29950
	3		29500
	4		29500
Common to two squares	5	2/	29850
	6	2/	29900
	7	2/	29850
	8	2/	29600
	9	2/	29500
	10	2/	29500
	11	2/	29500
	12	2/	29600
Common to four squares	13	4/	29600
	14	4/	29650
	15	4/	29700
Total number of levels	32		949200
Ave egl in the cut area	÷32		29663
SSL			29750
Concrete slab			(150)
Hardcore base			(100)
CUT			(29500)
			163

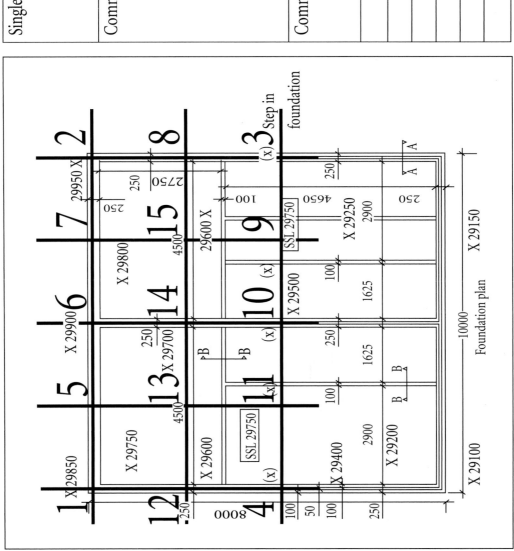

Diagram 6.2 Cut gridlines and calculations.

Single corners	1		29500
	2		29500
	3		29150
	4		29100
Common to two squares	5	2/	29500
	6	2/	29500
	7	2/	29500
	8	2/	29300
	9	2/	29150
	10	2/	29200
	11	2/	29150
	12	2/	29300
Common to four squares	13	4/	29300
	14	4/	29300
	15	4/	29250
Total number of levels	32		937850
Ave egl in the fill area	÷32		29308
SSL			29750
Concrete slab			(150)
Hardcore base			(100)
			(29500)
FILL			**(192)**

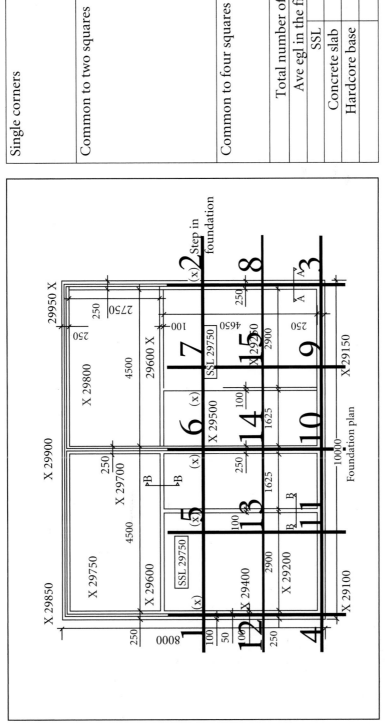

Diagram 6.3 Fill gridlines and calculations.

Table 6.3 Practical application: Pair of houses (*Continued*)

Substructure

<u>EXCAVATION</u>

			<u>Area</u>			
			10000 ✓	x		8000 ✓
			(500) ✓			
			<u>250</u> ✓			
			250 ✓			
			÷ 2 ✓			
			= <u>125</u>			
			Spread = 2/125 ✓ <u>250</u> ✓			<u>250</u> ✓
			<u>10250</u> ✓			<u>8250</u> ✓

<u>Depth</u>

	SSL 29750 ✓
Hardcore 150 ✓	
Concrete <u>100</u> ✓ (250) ✓	
Formation level <u>29500</u> ✓	

<u>Cut</u>

Ave egl	29663 ✓
Formation level	(29500) ✓
Cut	<u>163</u> ✓

<u>Maximum Cut</u>

Highest egl Say	29900 ✓
Formation level	(29500) ✓
Cut	<u>400</u> ✓

<u>Fill</u>

Ave egl	29308 ✓
Formation level	(29500) ✓
Fill	<u>192</u> ✓

10.25	Bulk excavation, ne	
8.25	2m deep, reduce levels	
.16	[5.6.1.1	
10.25		
4.00	[<u>Less</u> Fill area	*The area of the fill is deducted as no excavation will take place.*
.16		

Table 6.3 Practical application: Pair of houses (*Continued*)

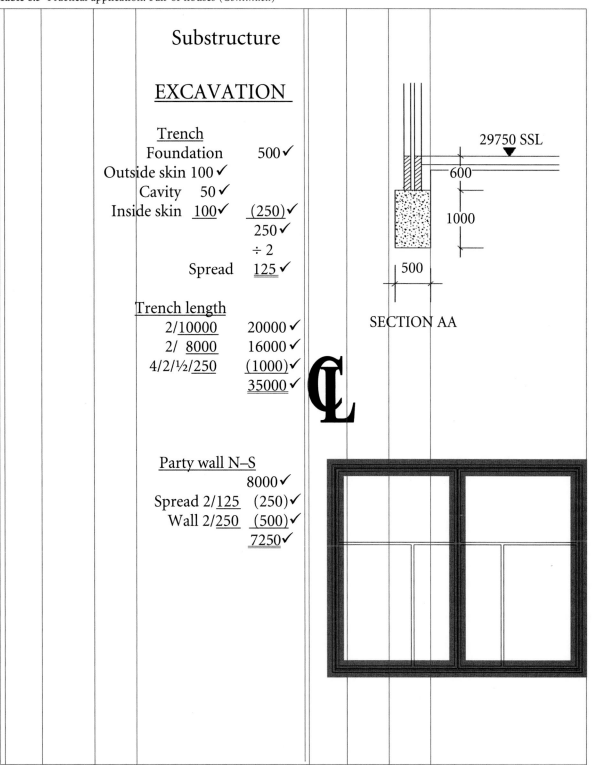

Substructure

EXCAVATION

Trench
Foundation   500 ✓
Outside skin 100 ✓
Cavity   50 ✓
Inside skin   100 ✓   (250) ✓
                           250 ✓
                           ÷ 2
Spread   125 ✓

Trench length
2/10000   20000 ✓
2/ 8000   16000 ✓
4/2/½/250   (1000) ✓
             35000 ✓

Party wall N–S
             8000 ✓
Spread 2/125   (250) ✓
Wall 2/250   (500) ✓
             7250 ✓

29750 SSL
600
1000
500

SECTION AA

Table 6.3 Practical application: Pair of houses (*Continued*)

Substructure

Internal walls E –W

(NB Not 4500)   2/4625   9250 ✓
Spread 4/125   (500) ✓
8750 ✓

The dimension that was incorrectly shown as 4500 on the drawing has been corrected to 4625 in this calculation.

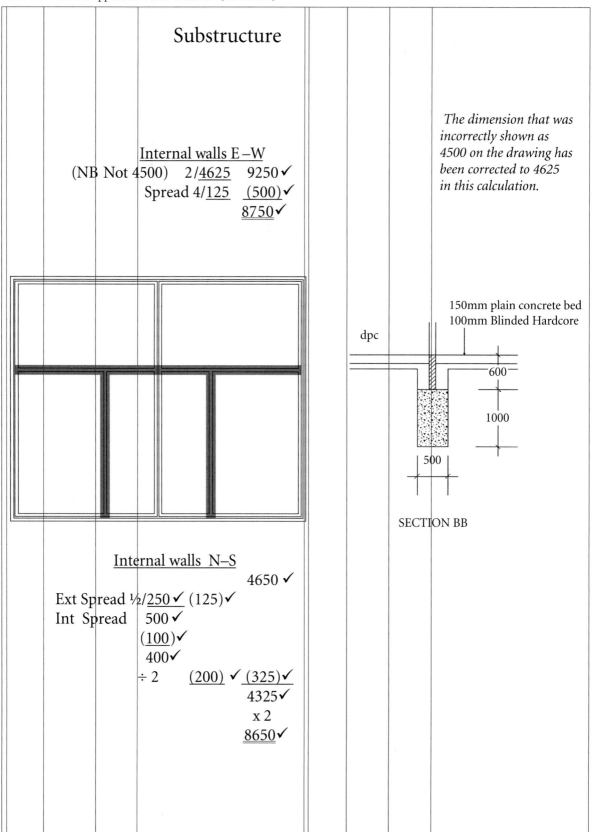

150mm plain concrete bed
100mm Blinded Hardcore

dpc

600

1000

500

SECTION BB

Internal walls N–S

4650 ✓
Ext Spread ½/250 ✓ (125) ✓
Int Spread   500 ✓
(100) ✓
400 ✓
÷ 2   (200) ✓ (325) ✓
4325 ✓
x 2
8650 ✓

Table 6.3 Practical application: Pair of houses (*Continued*)

Substructure

EXCAVATION

Trench Depth
Cut area

SSL	29750
Slab	100
Hardcore	150	(250)
Slab formation level	29500

Bkwk	600
Footings	1000	(1600)
Trench formation level	27900

Cut area egl   29663
(27900)
1763
Less RL (163)
1600

29750 SSL

600

1000

500

SECTION AA

Additional fill to the step

Ave egl	29308

SSL	29750
Slab	100
Hardcore	150
Bkwk	600
Footings	1000
Step	450 (2300) (27450)

1858

450

500

450

STEP IN FOUNDATION

Additional excavation in fill

1858
(1600)
258

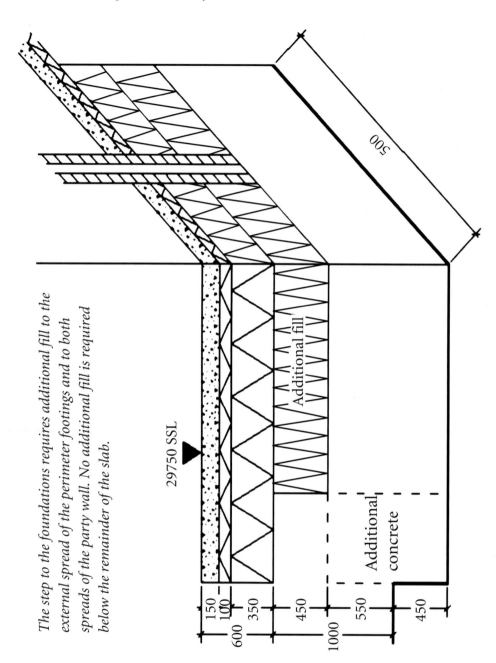

The step to the foundations requires additional fill to the external spread of the perimeter footings and to both spreads of the party wall. No additional fill is required below the remainder of the slab.

29750 SSL

500

150
100
600
350
450
1000
550
450

Additional fill

Additional concrete

Diagram 6.4 Step in foundations.

Table 6.3 Practical application: Pair of houses (*Continued*)   83

Substructure

	Foundation excavation, ne 2m deep, trenches [5.6.2.1		In-situ concrete, grade c25, horizontal work, > 300 thick, in structures, foundations [11.2.2.2.1
35.00 .50 1.60	[Perimeter wall	35.00 .50 1.00	[Perimeter wall
7.25 .50 1.60	[Party wall	7.25 .50 1.00	[Party wall
8.75 .50 1.60	[Int walls E–W	8.75 .50 1.00	[Int walls E–W
8.65 .50 1.60	[Int walls N–S	8.65 .50 1.00	[Int walls N–S
5/ 4.00 .50 .25	Say ½/8000 = 4000 [N–S in fill	5/ .50 .50 .45	[Steps
10.00 .50 .25	[E–W in fill		
		5/ .50	Plain formwork, sides of fnds, < 500 high [11.13.1 [Steps

Table 6.3 Practical application: Pair of houses (*Continued*)

Substructure

Facings to perimeter	
2/10000	20000
2/8000	16000
4/2/½/100	(400)
	35600

Brickwork

Party wall	8000
Cavity at perimeter 2/250	(500)
	7500

Height Say 4 courses 4/75 300

Facings to bottom of slope

	10000
2/½/8000	8000
	18000

Walls, 100mm th, blockwork, dense concrete blocks, _____ N/mm², c/m, (1:4) skins of hollow walls
[14.1.2.1

Height
EGL at step 29500
Lowest egl (29100)
400

2/ 35.00	
.60	
2/ 7.50	[Perimeter wall
.60	
2/2/ 4.65	[Party wall
.60	
2/ 10.00	[Int walls
.45	
2/ 4.00	[Extra for fill areas
.45	
3/2/ 4.00	
.45	

35.00	Ddt
.30	Walls, 100mm th, blockwork, dense
½/ 18.00	concrete blocks,
.40	_____ N/mm², c/m, (1:4)
	[14.1.2.1

&

Add
Walls, ½ b. th. bkwk, facework, PC (£500/1000), pointing in g/m (1:1:6)
[14.1.1.1.

2/ 35.00		Forming cavities, 50mm wide, s/s ties @ 5/m²
.60	42.00	
2/ 7.50		[14.14.1.1
.60	9.00	
3/ 4.65		&
.45	6.28	Cavity fill, plain concrete, (1:10)
10.00		
.45	4.50	61.78 m² x .05 = 3.09 m³
	61.78	

2/ 35.00	DPC, <300 wide, lead cored, 1.8mm thick, Code 4, horizontal, bedded in c/m (1:3)
2/ 7.50	
2/2/ 4.65	[14.16.1.3

Table 6.3 Practical application: Pair of houses (*Continued*)

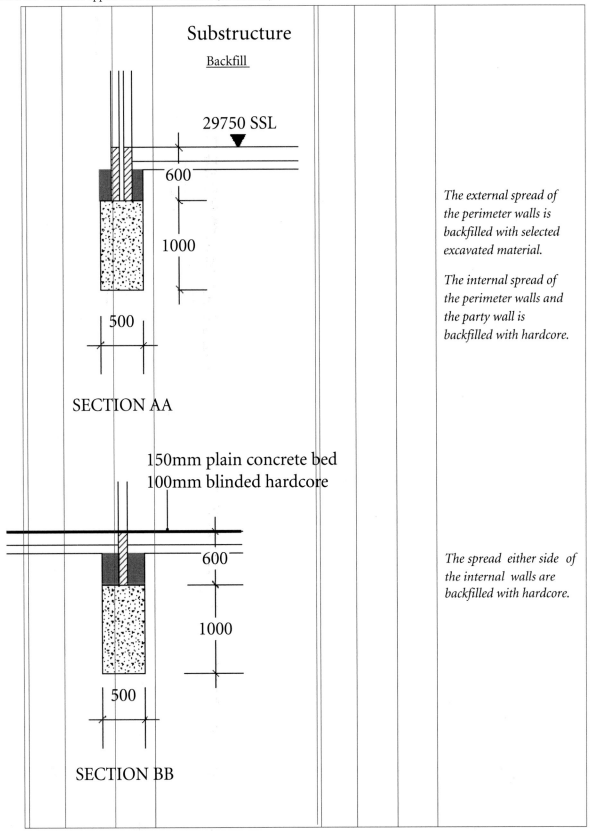

Substructure

Backfill

29750 SSL

600

1000

500

SECTION AA

150mm plain concrete bed
100mm blinded hardcore

600

1000

500

SECTION BB

The external spread of the perimeter walls is backfilled with selected excavated material.

The internal spread of the perimeter walls and the party wall is backfilled with hardcore.

The spread either side of the internal walls are backfilled with hardcore.

Table 6.3 Practical application: Pair of houses (*Continued*)

Substructure

		Dimensions	Description
	Backfill	36.50 .13 .35	Filling obtained from selected excavated material, <500 deep, 350 deep [5.11.1.1.2 [External spread
	Centre line for external spread		
	2/10000 20000 2/8000 16000 4/2/½/125 500 36500 ℄		
	Centre line for internal spread		
	2/10000 20000 2/8000 16000 Less wall thickness 4/2/250 (1000) Internal perimeter 35000 4/2/½/125 (500) 34500 ℄	34.50 .13 .35	Imported filling, hardcore, <500 deep, 350mm thick, compacting in 100mm layers [5.12.2.1 [Internal spread
	Depth for external & internal spread 600 (150) (100) 350	2/ 7.25 .13 .35	[Party walls
		2/ 8.75 .20 .35	[N–S
	Spread for internal walls Foundation 500 Blockwork (100) 400 ÷ 2 Spread 200	2/ 8.65 .20 .35	[E–W
		2/5/4.00 .20 .45	Imported filling, hardcore, <500 deep, 450mm thick, compacting in 100mm layers [5.12.2.2 [Step
		2/10.00 .20 .45	

Table 6.3 Practical application: Pair of houses (*Continued*)

Substructure

Reinforced concrete, grade C25, horizontal work, < 300 thick, beds
[11.2.1.2.1
67.51 m² x .15 = 10.13 m³

	9.50	
Ddt	7.50	71.25
7.50		
.25		(1.88)
2/2/4.65		
.10		(1.86)
		67.51

[Party wall

[Internal walls

&

Imported filling to excavation, hardcore, <500, ave 100mm th
[5.12.2.1
67.51 m² x .10 = 6.75 m³

&

Imported filling, sand blinding bed, ne 50mm th, obtained offsite, level to falls & cross falls
[5.12.1.1
67.51 m² x .05 = 3.38 m³

&

Surface finishes to reinforced concrete, power floating, to top surfaces, horizontal
[11.9.1

&

Mesh, A142, 2.22kg/m²
[5.37.1
67.51 m² x 2 = 135.02 m²

	9.50	
	4.00	
Ddt	.26	
4.00		
.25		
.26		
2/4.00		
.10		
.26		

Slab

	10000	8000
2/250	(500)	(500)
	9500	x 7500

÷ ½

= 3750

Say 4000

Filling to fill area
Imported filling to excavation, hardcore, <500, ave 258mm th
[5.12.2.2

[Party wall

[Internal walls

End of sloping site

6.4   SELF-ASSESSMENT EXERCISE: GRID OF LEVELS

Prepare a grid of levels using the plan and section on Drawings SDCO/2/E6/1, SDCO/2/E6/2 and SDCO/2/E6/3 in Appendix 6. Please prepare your own measurement using blank double dimension paper in Appendix 1a and also prepare a query sheet of problems that you have encountered. Compare your own work with the proposed solution included in Appendix 6 (Table E6.1) and self-assess your work on the assessment sheet in Appendix 1b.

To provide further assistance there are dedicated websites at http://ostrowski quantities.com and at Wiley Blackwell (http://www.wiley.com/go/ostrowski/measurement). It is hoped that the provision of this will go some way towards explaining the concepts and principles more clearly than using the printed word alone.

7 Underpinning

Measurement Using the New Rules of Measurement, First Edition. Sean D.C. Ostrowski.
© 2013 John Wiley & Sons, Ltd. Published 2013 by John Wiley & Sons, Ltd.

7.1   MEASUREMENT INFORMATION

Drawings

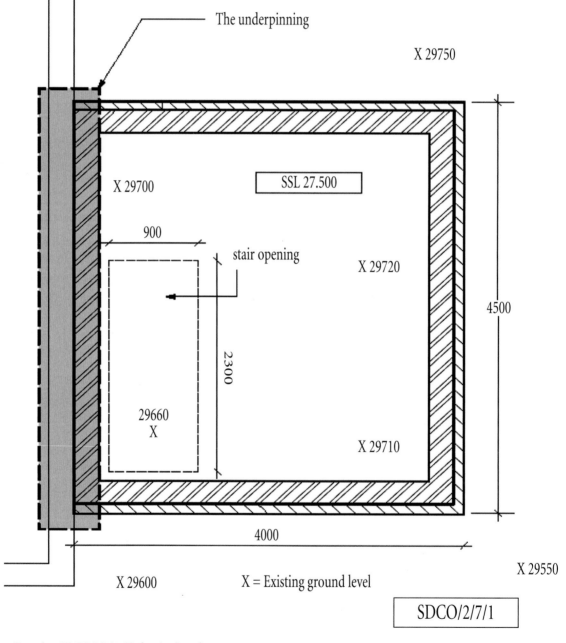

The underpinning

X 29750

X 29700

SSL 27.500

900

stair opening

X 29720

4500

2300

29660
X

X 29710

4000

X 29600   X = Existing ground level

X 29550

SDCO/2/7/1

Drawing SDCO/2/7/1 Underpinning plan.

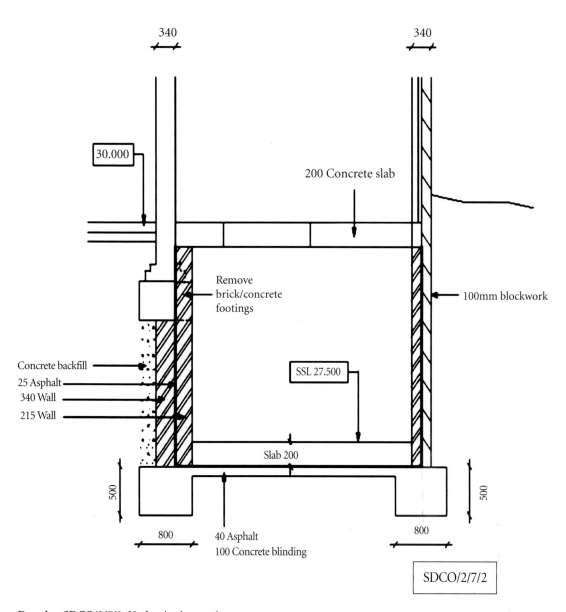

340

340

30.000

200 Concrete slab

Remove
brick/concrete
footings

100mm blockwork

Concrete backfill

25 Asphalt

340 Wall

215 Wall

SSL 27.500

Slab 200

500

500

800

800

40 Asphalt
100 Concrete blinding

SDCO/2/7/2

Drawing SDCO/2/7/2 Underpinning section.

Specification

Table 7.1 Underpinning specification.

Specification S7 Underpinning
Precontract water level determined at 28.500.
All surplus excavated material to be removed from site.
Hardcore to be broken brick or stone blinded with 50mm concrete.
Mass concrete foundations and blinding beds to BS12, designed mix, grade C25, sulphate resisting cement, 40mm aggregate.
Basement walls to be Class B engineering bricks, SHP Warnham bricks in cement mortar (1:4), 215mm thick.
Protective external cladding to be 100mm dense concrete blocks ($7N/mm^2$) in cement mortar (1:3), built against asphalt tanking.
Temporary propping as described by Structural Engineer/at the Contractor's discretion.

Query sheet

Table 7.2 Underpinning query sheet.

QUERY SHEET	
BASEMENT UNDERPINNING	
QUERY (From the QS)	ANSWER (From the Architect/Engineer) (Assumptions are to be confirmed by QS)
1. Spoil to be removed from site	Confirmed xx.xx.11
2. Underpinning	Assumed. Confirmed xx.xx.xx
3. Scale of Section	Assumed. Confirmed xx.x.xx
4. Working space	Yes. Confirmed xx.xx.xx
5. Backfill to be mass concrete.	Yes. Confirmed xx.xx.xx
6. Specification for Warnham bricks	PC £500/1000
7. Assume stretcher bond	Confirmed xx.xx.xx
8. DPC specification	Assumed. Confirmed xx.xx.xx
	etc.

7.2 TECHNOLOGY

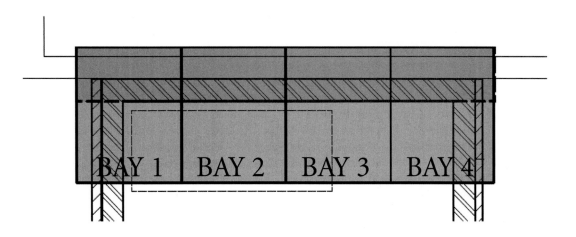

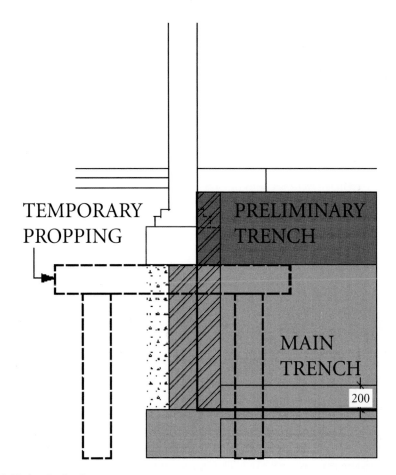

Diagram 7.1 Underpinning bays.

The basement is being excavated immediately adjacent to an existing building. This requires underpinning to the adjacent building before the construction work to the basement is undertaken.

Preliminary trench

A preliminary trench is necessary to expose the extent of the existing foundations. This will establish the depth and width of the existing foundations. The length of the trench will extend beyond the foundations to allow access at both ends, say 600 in this case. The width of the trench will be sufficient to allow access for working space, say 2000 beyond the external face of the external wall. The depth corresponds to the foundations for the existing building. The preliminary trench is excavated as a single length to expose the existing foundations and provide guidance on the underpinning procedures. The surplus spoil is removed and the trench requires earthwork support and levelling and compacting. While excavating the preliminary trench the exposed part of the existing foundations is to be broken up and removed.

Underpinning pits

The main trench excavates the access below the existing foundations. The length is the same as the preliminary trench. The total width is the width of the new foundations plus sufficient additional width to allow access for working space. The allowance is the same as that for the preliminary trench. The depth of the main trench extends down to the base of the new foundations for the basement. (The basement is measured in Chapter 5.)

Earthwork support

The preliminary trench and underpinning pits require earthwork support. The main trench removes the support for the foundations for the existing building. It is therefore completed in alternate sections or bays to ensure that only a limited length of the foundations is exposed at any one time. Temporary shoring is necessary to support the particular section of the foundation that has been exposed and will extend beyond the width of the existing foundations. In addition each bay has earthwork support on all four sides.

Alternate bays

The concrete backfill, the concrete footings, the formwork and the brickwork foundations are also constructed in these alternative bays. The existing precontract water level is below the datum of the slab so allowances will need to be included for ground water in addition to surface water.

Measurement

The NRM represents a simplified measure of the actual work involved. It measures the concrete foundations and brickwork wall as linear metres with brief descriptions. To be comprehensive the measurement and pricing need to reflect the total work involved. This can be done by building up a compound rate. The worked example in Table 7.3 shows the measurement using NRM 2. This is followed by a comprehensive measurement of all the work necessary in order to build up a compound rate.

Table 7.3 Practical application: Underpinning to basement.

			Substructure BASEMENT UNDERPINNING EXISTING BUILDING Drawings SDCO/2/7/1 Plan SDCO/2/7/2 Section Specification S7 SDCO May 2012				*The title page includes the trade name, the name of the contract, the full drawing schedule with dated revisions, the dated specification, the name of the measurer and the date.*

Table 7.3 Practical application: Underpinning to basement (*Continued*)

Basement underpinning

DEPTH

Average existing gl	29700
	29660
	59360
	÷ 2
Ave egl =	29680

	29680		
SSL	27500		
Slab	200		
Asphalt	40		
Fndn	500	(740)	(26760)
EXCAVATION			2920

4.80   4.80

LENGTH

Scaled	2/150	300
		4500
		4800

WIDTH

Assumed width beyond the face of the foundation wall to allow sufficient working space, the allowance.

Footings 800	
Less Bkwk (215)	585
Allowance Say	1500
	2085

The side casts are to provide dimensions for the depth and maximum width as NRM 8.1.1.1.

Underpinning, foundations, concrete Grade C20, 40mm aggregate, approximately 5000 long x 2100 maximum width x 3000 deep, comprising in-situ concrete footings 800 x 500, in 4 nr bays, each 1200 wide
                                    [8.1.1.1

&

Ditto, walls, 340 thick x 1450 high, brickwork, SHP Warnham bricks in c/m (1:4) built against other work, keyed one side, bonding to other work, building overhand, in bays, abd
                                    [8.1.2.1

End of basement underpinning

Table 7.3 Practical application: Underpinning to basement (*Continued*)

Substructure

To obtain an accurate price for the work it is necessary to measure and price all the work and then combine these items into the compound item in accordance with the NRM 2descriptions. This is similar to the measurement process necessary to measure estimate and cost plans in accordance with NRM 1.

All the items in the following measurement are preamble. They provide significant information concerning the nature of the work and will assist in the pricing of the work. Preambles do not normally generate any cost but in this case there are two cost significant items.

The first is the 'temporary support for the existing structures'.

The second cost item is the work being carried out in sections. This requires individual bays for all the work. In this case there are 4 bays and the work in each will have to be carried out 4 times.

The work includes the breaking up of the existing ground, preliminary trench excavation to the bottom of the existing foundations, disposal, earthwork support, the shoring and the scaffolding etc.

Table 7.3 Practical application: Underpinning to basement (*Continued*)

Basement Underpinning

<u>EXCAVATION</u>

<u>4</u>	*Trial pits*			The information concerning the method of construction is provided by the structural engineer.
Item	*Underpinning to location and extent shown on the drawings in alternate 1200 sections to a total length of 4800 and 4 sections in all*			The additional work and pricing that is necessary to proceed in alternate bays and the temporary support work is included in the compound rate for the work. The method of compounding costs is set out at the end of the measurement. All the dimensions are shown in italics and are not measureable in the NRM for underpinning but are included in the compound description.
Item	*Temporary support for existing structures*			

Table 7.3 Practical application: Underpinning to basement (*Continued*)

Basement underpinning

EXCAVATION

These side casts are to calculate the dimensions of the preliminary and main trenches.

PRELIMINARY TRENCH

	DEPTH
Average existing gl	29700
	29660
	59360
	÷ 2
Ave egl =	29680

	29680
Existing slab level 30000	
To base of extg footings Scaled	(1100)
	(28900)
EXCAVATION	780

Precontract gwl	28500
Formation level	(27360)
Excavation below gwl	1140

WIDTH

Assumed width beyond the face of the foundation wall to allow access to existing foundations.
1500

LENGTH

Allowance for access at both ends Say	
2/600	1200
	4500
	5700

MAIN TRENCH IN BAYS

	DEPTH
Average existing gl	29700
	29660
	59360
	÷ 2
Ave egl =	29680
	29680

	SSL 27500		
Slab	200		
Asphalt	40		
Fndn	500	(740)	(26760)
	EXCAVATION		2920

Precontract gwl	28500
Formation level	(27360)
Excavation below gwl	1140

WIDTH

Assumed width beyond the face of the foundation wall to allow sufficient working space.
1500

Footings 800	
Less Bkwk (215)	585
	2085

LENGTH

Allowance for access at both ends Say	
2/600	1200
	4500
	5700

Table 7.3 Practical application: Underpinning to basement (*Continued*)

Basement underpinning

5.70 1.50 .78	6.67	*Foundation excavation, ne 2.00m deep, preliminary trenches* [5.6.2.1 *[Preliminary trench*				*All the dimensions in italics are not measureable in the NRM but are included in the compound description.*
Item		*Disposal of excavated material off site* [5.9.2				*Excavation and cart away for the preliminary trench and the individual underpinning pits are separate operations.*
5.70 2.09 2.92	34.79	*Foundation excavation, 2–4m deep, pits, in bays 1200 wide* [5.6.2.2 *[Underpinning pits*				*The depth below the existing ground water level is to be measured where available.* *Where ground water exists the removal of it is included as an 'Item'.*
Item		*Disposal of excavated material off site* [5.9.2		5.70 .78 2/1.50 .78	6.79	*The planking and strutting for underpinning is complex.* *Earthwork support, maximum depth 780, distance between faces, >2m preliminary trench.* [5.8.1.1
5.70 2.09 1.14	13.58	*Extra over all types of excavation, excavating below gwl* [5.7.1.3		4/2/1.20 2.92 4/2/2.08 2.92	76.85	*Ditto, maximum depth 2650, underpinning pits, below gwl, next to existing building, left in, in alternate 1200 wide sections.* [5.8.1.1
Item		*Disposal of ground water, 1.00 below egwl* [5.9.1.1				

Table 7.3 Practical application: Underpinning to basement (*Continued*)

Basement Underpinning

EXCAVATION

REMOVE BRICK
CONCRETE
FOOTINGS

4.70			*Extra over all types of excavation irrespective of depth, breaking up, concrete, projecting foundation*	
.15				
.40	.28		[5.7.2.3	

$$3/75 = 225$$

			Ditto, masonry
4.70			[5.7.2.4
.15			
.23	.16		

					Plain in-situ concrete, Grade C20, 40mm aggregate, in bays, isolated, poured against earth
			5.70		
			.80		
			.50	2.28	
4/	1.20		*Surface treatment, compacting bottom of excavation*		[11.1.2.2.1
	2.09	10.03	[Bottom of each pit		

```
                              800
                            (340)
                              460
                              ÷ 2
              Spread =        230

Brickwork Width  Say         150

                    Depth
                    Scaled   400

                    Length
                             4500
           Scaled 2/100      200
                             4700
```

Table 7.3 Practical application: Underpinning to basement (*Continued*)

Basement underpinning

4/	1.20		Plain formwork, side of foundation, in bays, >500 high [11.13.1	4/	1.20 4.80	Underpinning, prepare underside of existing work to receive pinning up and new work, in bays abd, 585 wide [14.25.1
	.50					
4/2/	2.09					
	.50					
		10.76				&
						Wedging and pinning between foundations and brickwork, 340 wide [14.25.1
4/	1.20		Plain in-situ concrete, grade C20, 40mm aggregate, packed behind brick work, in bays, abd [11.1.1.1.1 [Bays [Ends			800
	.23					(215)
	1.45	1.60				585
2/	.60					Raking out the brickwork face of the new brickwork in the underpinning is to receive the asphalt tanking that is part of the basement. The basement is measured in Chapter 5.
	.23					
	1.45	.40				
		2.00				
				4.50		Rake out existing brickwork, 10mm joints, engineering bricks, to receive asphalt (measured separately) [14.21.1.1
				1.45	6.53	
	4.80		Walls, brickwork, 340 thick, SHP Warham bricks, c/m (1:4), vertical, built against, other work, bonding to other work, building overhand [14.1.1.4.1			The area beyond the face of the underpinning would normally be backfilled with hardcore. It is measured here to complete the underpinning section.
	1.45					
		6.96				
				5.70		Imported filling to excavation, hardcore > 500 deep, obtained off site, compacted in 250mm thick layers [5.12.3.1
				1.50		
				2.92		
					24.97	

The following table demonstrates how to build up a compound item for this complex underpinning work and enable it to be described in accordance with NRM 2 section 8 Underpinning.

Table 7.4 Underpinning compound measurement.

COMPOUND MEASUREMENT FOR UNDERPINNING				
Description	**Quantity**	**Rate**	**£**	
FOUNDATIONS				
Trial pits	4 nr	1,000.00	4,000.00	
Alternate sections	Item		5,000.00	
Temporary support	Item		10,000.00	
Excavate preliminary trench	7m³	25.00	175.00	
Disposal off site	7m³	10.00	70.00	
Excavate underpinning pits	35m³	100.00	3,500.00	
Disposal off site	35m³	10.00	3,500.00	
Excavate below egl	14m³	5.00	70.00	
Ground water	Item		100.00	
EWS 780 deep	7m²	10.00	70.00	
EWS 2650 deep	77m²	15.00	1,155.00	
BO concrete	1m³	50.00	50.00	
BO brickwork	1m³	100.00	100.00	
L & R	10m³	5.00	50.00	
Foundations	2m³	200.00	400.00	
Formwork	11m²	50.00	550.00	25,640.00
		Length of concrete foundations		÷ 5m
		Rate per metre		= 5,128/lm
NRM method of measurement				
Underpinning, foundations Ref 8.1	**5m**	**5,128/lm**	**£25,640.00**	

(Continued)

Table 7.4 (*Continued*)

COMPOUND MEASUREMENT FOR UNDERPINNING				
Description	Quantity	Rate	£	
BRICKWORK				
Brickwork	7m^2	250.00	1,750.00	
Concrete backfill	2m^3	300.00	600.00	
Prepare underside	5m	100.00	500.00	
Wedge & pin	5m	200.00	1,000.00	
Rake out brickwork	7m^2	5.00	35.00	
Hardcore backfill	25m^3	50.00	1,250.00	5,135.00
			Length of brickwork wall	÷ 5m
			Rate per metre	= 1,027/lm
NRM method of measurement				
Underpinning, Walls Ref 8.2	**5m**	**1,027/lm**	**£5,135.00**	

7.4   SELF-ASSESSMENT EXERCISE: TRENCH EXCAVATION

Calculate suitable bay widths and measure the earthwork support for the underpinning pits shown on Drawing SDCO/2/E7/1 in Appendix 7. Please prepare your own measurement using blank double dimension paper in Appendix 1a and also prepare a query sheet of problems that you have encountered. Compare your own work with the proposed solution included in Appendix 7 (Table E7.1) and self-assess your work on the assessment sheet in Appendix 1b.

To provide further assistance there are dedicated websites at http://ostrowski quantities.com and at Wiley Blackwell (http://www.wiley.com/go/ostrowski/ measurement). It is hoped that the provision of this will go some way towards explaining the concepts and principles more clearly than using the printed word alone.

8 Reinforced Concrete Frame

8.1 Measurement information
 • Drawings
 • Specification
 • Query sheet
8.2 Technology
8.3 Practical application: Reinforced concrete frame
8.4 Self-assessment exercise: Formwork

8.1 MEASUREMENT INFORMATION

Drawings

See Drawings SDCO/2/8/1 (reinforced concrete frame plan and details) and SDCO/2/8/2 (reinforced concrete frame section).

Measurement Using the New Rules of Measurement, First Edition. Sean D.C. Ostrowski.
© 2013 John Wiley & Sons, Ltd. Published 2013 by John Wiley & Sons, Ltd.

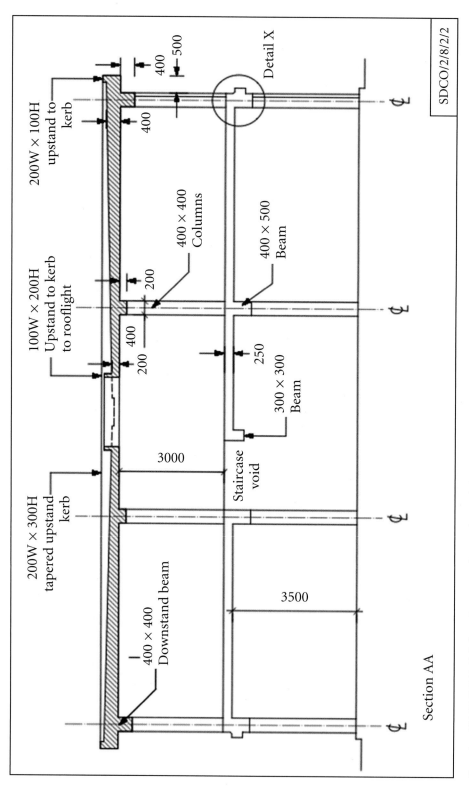

Drawing SDCO/2/8/1 Reinforced concrete frame plan and details.

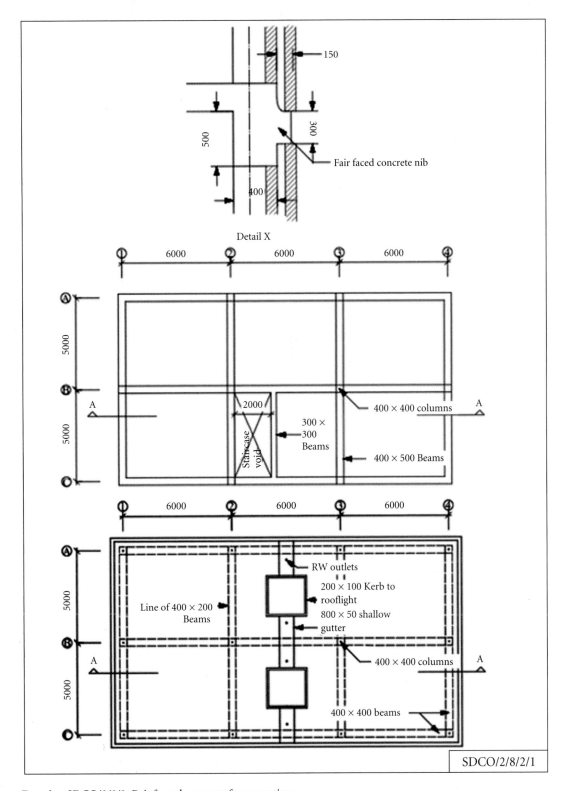

150

500

300

Fair faced concrete nib

400

Detail X

6000 6000 6000

5000

5000

A

2000

Staircase void

300 × 300 Beams

400 × 400 columns

400 × 500 Beams

A

6000 6000 6000

5000

5000

A

RW outlets

200 × 100 Kerb to rooflight

800 × 50 shallow gutter

Line of 400 × 200 Beams

400 × 400 columns

400 × 400 beams

A

SDCO/2/8/2/1

Drawing SDCO/2/8/2 Reinforced concrete frame section.

Specification

Table 8.1 Reinforced concrete frame specification.

Specification S 8 Reinforced Concrete Frame
Reinforced concrete columns to be ready mixed concrete, BS 5328, designated mix RC 30/37, 37mm aggregate.
Reinforced concrete beams to be designated mix RC 20/25, 25mm aggregate
Reinforced concrete roof and floor slabs to be designated mix 20/25, 25mm aggregate.
Plain formwork to all areas. Fair finish to face of concrete nib.
Reinforcement details to follow. Allow a provisional quantity of 2% by volume of concrete.

Query sheet

Table 8.2 Reinforced concrete frame query sheet.

QUERY SHEET	
RC FRAME	
QUERY (From the QS)	ANSWER (From the Architect/Engineer) (Assumptions are to be confirmed by QS)
1. Upper slopes of roof to be formed in concrete	Confirmed xx.xx.11
2. Full circumference of nib to be fair faced	Confirmed xx.xx.xx
3. Full circumference of nib to roof be fair faced	Confirmed xx.xx.xx
4. Nibs to S/C to be f/f	Confirmed xx.xx.xx
5. Reinforcement	Provisional 2% by volume of concrete. Confirmed xx.xx.xx

8.2   TECHNOLOGY

The building is a framework of reinforced concrete beams, columns and slabs.

The finished detail on the outside face of the nib at first floor level will require the formwork to provide a fair finish to the concrete and Detail X on the drawing in Table 8.3 provides the details.

The sloping top surface of the roof is difficult to achieve and may require an alternative method of forming the slope.

Table 8.3 Practical application: Reinforced concrete frame.

			Superstructure					
			REINFORCED CONCRETE FRAME					
			Drawings					
			SDCO/1/8/1 Plan					
			SDCO/1/8/2 Section					
			Specification S8					
								The title page includes the trade name, the name of the contract, the full drawing schedule with dated revisions, the dated specification, the name of the measurer and the date.
								The quantities are squared in this section because it is also used to demonstrate BQ preparation in Chapter 24.
			SDCO					
			May 2012					
			1					

Table 8.3 Practical application: Reinforced concrete frame (*Continued*)

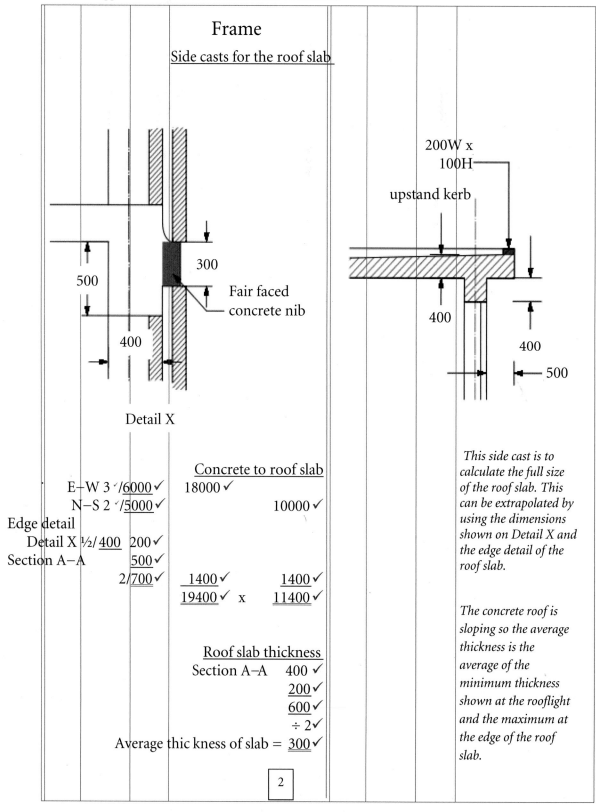

Frame

Side casts for the roof slab

300

500

Fair faced
concrete nib

400

Detail X

200W x
100H

upstand kerb

400

400

500

Concrete to roof slab

E–W 3 ✓/6000 ✓   18000 ✓
N–S 2 ✓/5000 ✓   10000 ✓
Edge detail
  Detail X ½/ 400   200 ✓
Section A–A   500 ✓
                  2/700 ✓   1400 ✓   1400 ✓
                        19400 ✓ x   11400 ✓

Roof slab thickness
Section A–A   400 ✓
              200 ✓
              600 ✓
              ÷ 2 ✓
Average thic kness of slab = 300 ✓

2

*This side cast is to
calculate the full size
of the roof slab. This
can be extrapolated by
using the dimensions
shown on Detail X and
the edge detail of the
roof slab.*

*The concrete roof is
sloping so the average
thickness is the
average of the
minimum thickness
shown at the rooflight
and the maximum at
the edge of the roof
slab.*

Table 8.3 Practical application: Reinforced concrete frame (*Continued*)

Frame

Side casts for the roof beams

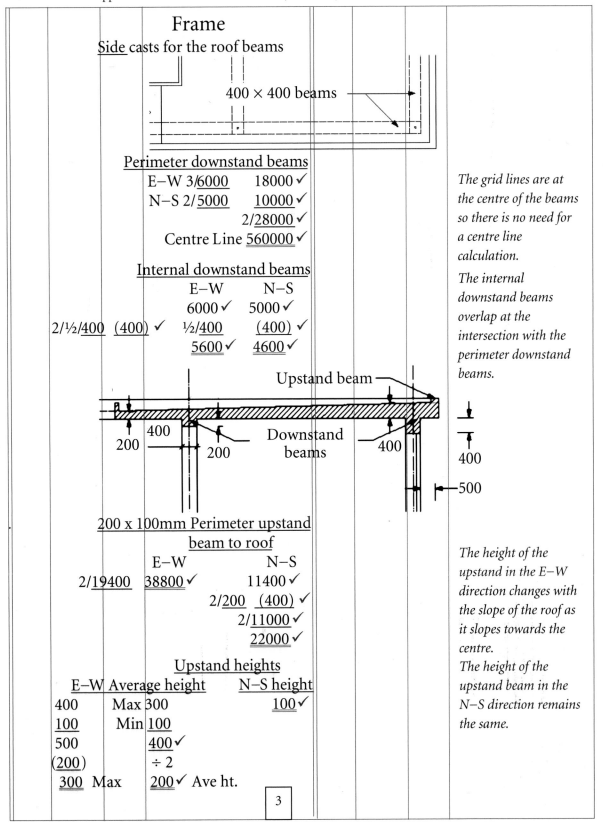

400×400 beams

Perimeter downstand beams

E–W 3/6000	18000 ✓
N–S 2/5000	10000 ✓
	2/28000 ✓
Centre Line	560000 ✓

The grid lines are at the centre of the beams so there is no need for a centre line calculation.

Internal downstand beams

	E–W	N–S
	6000 ✓	5000 ✓
2/½/400 (400) ✓	½/400	(400) ✓
	5600 ✓	4600 ✓

The internal downstand beams overlap at the intersection with the perimeter downstand beams.

Upstand beam

400

200   200   Downstand   400
  400   beams

500

200 x 100mm Perimeter upstand beam to roof

	E–W	N–S
2/19400	38800 ✓	11400 ✓
		2/200 (400) ✓
		2/11000 ✓
		22000 ✓

The height of the upstand in the E–W direction changes with the slope of the roof as it slopes towards the centre.

The height of the upstand beam in the N–S direction remains the same.

Upstand heights

E–W	Average height	N–S height
400	Max 300	100 ✓
100	Min 100	
500	400 ✓	
(200)	÷ 2	
300 Max	200 ✓ Ave ht.	

3

Table 8.3 Practical application: Reinforced concrete frame (*Continued*)

Frame

Side casts for the roof lights beams

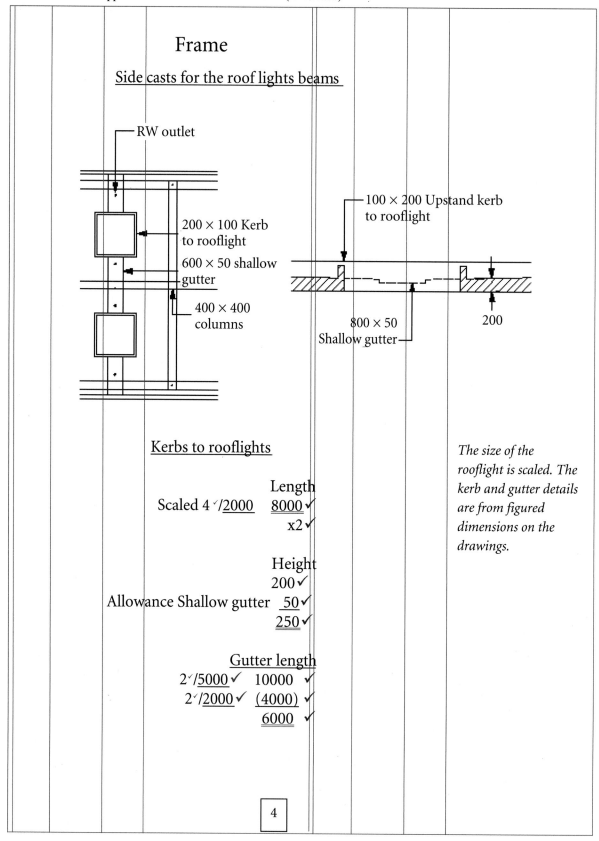

— RW outlet

200 × 100 Kerb
to rooflight

600 × 50 shallow
gutter

400 × 400
columns

— 100 × 200 Upstand kerb
to rooflight

800 × 50
Shallow gutter—

200

Kerbs to rooflights

		Length
Scaled 4 ✓/2000		8000 ✓
		x2 ✓
		Height
		200 ✓
Allowance Shallow gutter		50 ✓
		250 ✓

		Gutter length
2 ✓/5000 ✓	10000 ✓
2 ✓/2000 ✓	(4000) ✓
	6000 ✓

The size of the rooflight is scaled. The kerb and gutter details are from figured dimensions on the drawings.

4

Table 8.3 Practical application: Reinforced concrete frame (*Continued*)

Frame

NRM 2 p. 151, column
4, item 3 voids > 0.05m³
are to be deducted.

RC to Roof

Reinforced in-situ
concrete, ready mixed
concrete, BS 5328,
designated mix RC
20/25, horizontal
work, slabs, thickness
ne 300 thick, in
structures, in bays ave.
26m²
  [11.2.1.2.*.3
  [Ddt Rooflights

  [Ddt Gutter

19.40 ✓		
11.40 ✓		
.30 ✓	66.35 ✓	
Ddt		
2/2.00 ✓		
2.00 ✓		
.20 ✓	(1.60) ✓	
6.00 ✓		
.80 ✓		
.05 ✓	(.24) ✓	
	64.51 ✓	

	Rooflight	Gutter
	2.00 ✓	6.00 ✓
	2.00 ✓	.80 ✓
	.20 ✓	.05 ✓
	.80 ✓	.24 ✓

*The thickness of the slab
at the rooflights is 200mm .*

2/19400	38800
	11400
2/200	(400)
2/11000	22000

RC ditto, beams,
thickness > 300 thick,
in structures
  [11.2.2.2
  [Perimeter beams

56.00 ✓	
.40 ✓	
.40 ✓	8.96 ✓

RC ditto, thickness
ne 300 thick, in
structures, beams
  [11.2.1.2
  [Internal beams

[Top of intermediate
columns

[Upstands E−W

[Upstands N−S

[Kerbs

3/ 5.60 ✓		
.40 ✓		
.20 ✓	1.34 ✓	
4/ 4.60 ✓		
.40 ✓		
20 ✓	1.47 ✓	
2/ .40 ✓		
.40 ✓		
20 ✓	0.06 ✓	
38.80 ✓		
.20 ✓		
20 ✓	1.55 ✓	
22.00 ✓		
.20 ✓		
10 ✓	.44 ✓	
2/ 8.00 ✓		
.10 ✓		
25 ✓	.40 ✓	
	5.26 ✓	

5

A — A

Table 8.3 Practical application: Reinforced concrete frame (*Continued*)

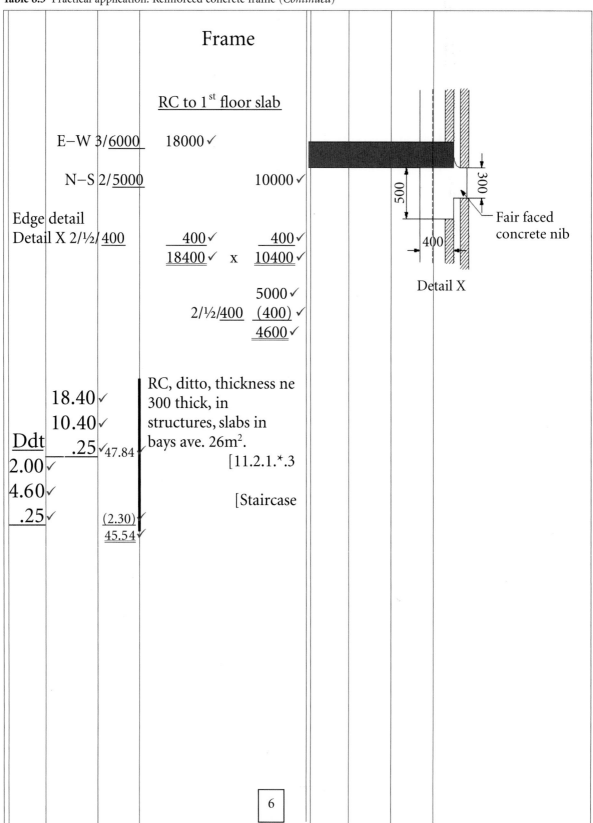

			Frame
			<u>RC to 1st floor slab</u>
	E–W 3/6000		18000✓
	N–S 2/5000		10000✓
Edge detail Detail X 2/½/400			400✓ 400✓ 18400✓ x 10400✓
			5000✓ 2/½/400 (400)✓ 4600✓
	18.40✓ 10.40✓ .25✓	47.84✓	RC, ditto, thickness ne 300 thick, in structures, slabs in bays ave. $26m^2$. [11.2.1.*.3
Ddt 2.00✓ 4.60✓ .25✓		(2.30)✓ 45.54✓	[Staircase

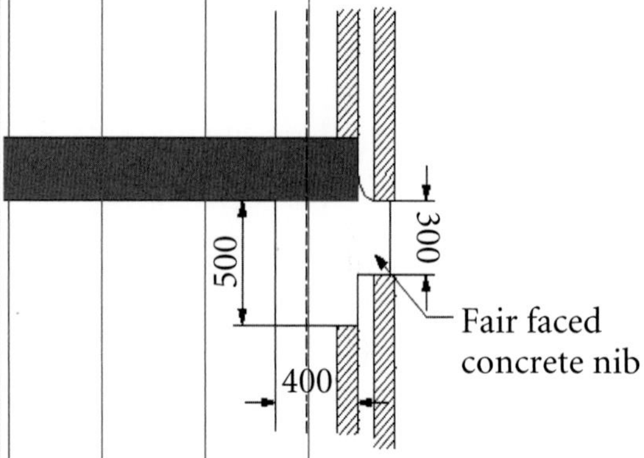

500
300
400
Fair faced
concrete nib

Detail X

6

Table 8.3 Practical application: Reinforced concrete frame (*Continued*)

Frame

RC beams to 1st floor slab

Nib 2/18400   36800 ✓
2/10400   20800 ✓   57600 ✓
4/2/½/150   600 ✓
58200 ✓

RC abd designated mix RC20/25, superstructure, horizontal work, beams, thickness > 300 thick, in structures

[11.2.2.2
[Beams E−W

[Beams N−S

3/ 18.00 ✓
.40 ✓
.50 ✓ 10.80 ✓
4/ 10.00 ✓
.40 ✓
.50 ✓ 8.00 ✓ [Intersections

Ddt
2/ .40 ✓
.40 ✓
.50 ✓ (.32) ✓ [S/C trimmer
2/1½/.40 ✓ 4.60 ✓
.40 ✓ .30 ✓ (.08) ✓ [Nib to detail X
.50 ✓ .30 ✓ .41 ✓
58.20 ✓
.15 ✓
.30 ✓ 2.62 ✓
21.43 ✓

The beams to the roof were measured net with the intersections added.

The beams to the first floor are measured gross with the intersections deducted.

Detail X

7

Table 8.3 Practical application: Reinforced concrete frame (*Continued*)

Frame

Columns

GF	FF	Internal
3500 ✓	3000 ✓	3000 ✓
(500) ✓	(400) ✓	(200) ✓
3000 ✓	2600 ✓	2800 ✓

12 / 3.00 ✓		RC abd designated
.40 ✓		mix RC 30/37,
.40 ✓	5.76 ✓	superstructure, vertical
10 / 2.60 ✓		work, 30N/mm²,
.40 ✓		columns, thickness >
.40 ✓	4.16 ✓	300 thick, in structures
2 / 2.80 ✓		[11.5.2.1.2
.40 ✓		
.40 ✓	.90 ✓	
	10.82 ✓	**1ˢᵗ floor finishes**

The internal columns between the first floor and the roof are longer because the downstand beams above them are 400 x 200 and those at the perimeter are 400 x 400.

18.40 ✓		Surface finishes,
Ddt 10.40 ✓		power floating, to top
2.00 ✓	191.36 ✓	surfaces [11.9.1
4.60 ✓		[Ddt S/C
	(9.20) ✓	
	182.16 ✓	

Tamping is now included (Ref. 9.*.*.7). Power floating is assumed and measurable.

Roof finishes

19400 ✓	11400 ✓	
2/200 ✓ (400) ✓	(400) ✓	
19000 ✓	11000 ✓	

Slope 200 high ÷ 9000 length
= 2.2%/1:45/c3°

19.00 ✓		Plain formwork,
Ddt 11.00 ✓		sloping top surface,
2/2.20 ✓	209.00 ✓	<15°
2.20 ✓		[11.28.1
	(9.68) ✓	
	199.32 ✓	[Ddt Rooflights

Forming the sloping top surface to the concrete roof is a significant labour item.

8

Table 8.3 Practical application: Reinforced concrete frame (*Continued*)

Frame

Formwork to soffit of roof

Roof	6000 ✓	5000 ✓	
2/1½/400 ✓	(400) ✓	(400) ✓	
	5600 ✓	4600 ✓	

½/6/	5.60 ✓		Formwork, soffits of horizontal work, for concrete < 300 thick, propping ne 300 high [11.15.1.1
	4.60 ✓ 77.28 ✓		

½/6/	5.60 ✓		Ditto 300–450 thick, do [11.15.2.1
	4.60 ✓ 77.28 ✓		

Formwork to roof downstand beams

		2/200 ✓	400 ✓
			400 ✓
3/400 ✓	1200 ✓		800 ✓

Formwork, sides and soffits of attached beams (17 nr), propping , < 3.00m high
[11.18.1.1
[Perimeter

[E–W

[N–S

[Intersections

	56.00 ✓	
	1.20 ✓ 67.20 ✓	
3/	5.60 ✓	
	.80 ✓ 13.44 ✓	
4/	4.60 ✓	
	.80 ✓ 14.72 ✓	
2/4/	.40 ✓	
	.30 ✓ .96 ✓	
	96.32 ✓	

9

The sloping thickness of the roof means the formwork is split between items 11.15.1.1 and 11.15.2.1.

The bays that are the soffits of the roof can be halved to represent that half of the roof will be above 300 mm thick and half will be below 300mm thick.

No deductions are made for voids <5.00m^2 (Item 6 of the notes. The rooflight void is 2.00 x 2.00 = 4m^2.

400 × 200 Downstand beam
400 × 400 Downstand beam

Add the sides and soffits of each beam first to calculate the total girth of each beam.

Adding the total number of beams of each girth enables the icing to reflect the number of uses possible for the formwork.

Adjustments for corners and intersections because gross lengths are used.

Table 8.3 Practical application: Reinforced concrete frame (*Continued*)

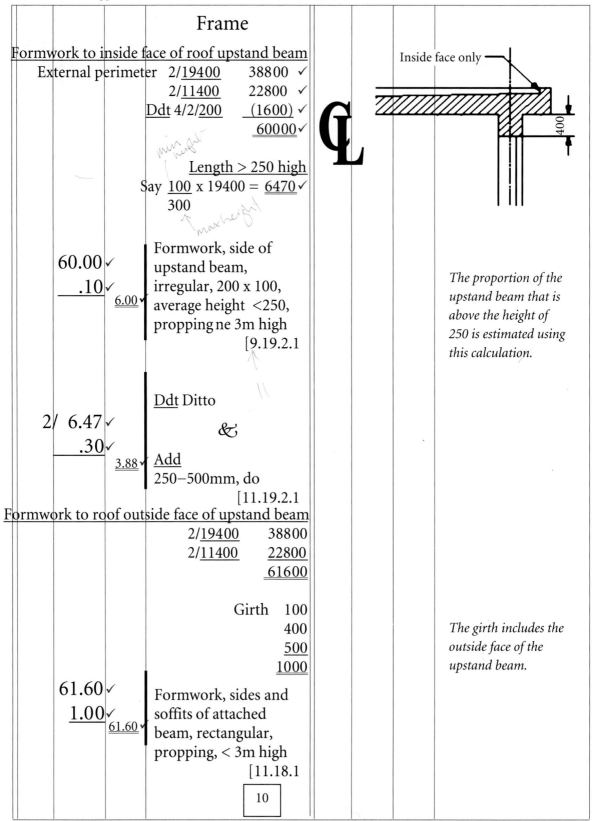

Frame

<u>Formwork to inside face of roof upstand beam</u>

External perimeter	2/<u>19400</u>	38800 ✓
	2/<u>11400</u>	22800 ✓
Ddt 4/2/<u>200</u>		(1600) ✓
		<u>60000</u> ✓

min height

<u>Length > 250 high</u>

Say $\dfrac{100}{300}$ x 19400 = <u>6470</u> ✓

max height

60.00 ✓		Formwork, side of
.10 ✓		upstand beam,
	<u>6.00</u> ✓	irregular, 200 x 100,
		average height <250,
		propping ne 3m high
		[9.19.2.1

		Ddt Ditto
2/ 6.47 ✓		*&*
.30 ✓		
	<u>3.88</u> ✓	<u>Add</u>
		250–500mm, do
		[11.19.2.1

<u>Formwork to roof outside face of upstand beam</u>

	2/<u>19400</u>	38800
	2/<u>11400</u>	22800
		<u>61600</u>

Girth	100
	400
	<u>500</u>
	<u>1000</u>

61.60 ✓		Formwork, sides and
1.00 ✓		soffits of attached
	<u>61.60</u> ✓	beam, rectangular,
		propping, < 3m high
		[11.18.1

10

Inside face only

400

The proportion of the upstand beam that is above the height of 250 is estimated using this calculation.

The girth includes the outside face of the upstand beam.

Table 8.3 Practical application: Reinforced concrete frame (*Continued*)

Frame

Formwork to upstand kerb to roof light

4/2000	8000 ✓
4/2/100	800 ✓
	8800 ✓

Girth 3/200 600 ✓

2/ 8.80 ✓
.60 ✓
10.56 ✓

Formwork, sides of upstand beams, 100 x 200, propping < 3m high
[11.19.1.1

Formwork to shallow gutter

2	4/5000	10000 ✓
	2/½/400	400 ✓
	2/500	1000 ✓
		11400 ✓

Rooflight 2/2000 4000 ✓
Kerb   2/2/100   400 ✓
Upstands 2/200   400 ✓   (4800) ✓
                          6600 ✓

6.60 ✓ 6.60 ✓

Formwork, recesses, (3nr) shallow gutter 800 x 50, to top of concrete slab
[11.31.1

4 ✓   4 ✓

Formwork, holes, diameter < 500, depth < 250, circular
[11.31.1

100 × 200 Upstand kerb to rooflight

800 × 50 Shallow gutter

200

11

Table 8.3 Practical application: Reinforced concrete frame (*Continued*)

Frame

Formwork to soffit of upper floor

	5600	✓
Staircase	(2 000)	✓
Beam	(300)	✓
	3300	✓

5/	5.60	✓	Formwork, soffits of horizontal work, for concrete < 300 thick, propping 3–4.5m high [11.15.1.2
	4.60	28.80 ✓	
	3.30	✓	
	4.60	15.18 ✓	
		141.98 ✓	

The bay that includes the staircase is adjusted.

The staircase opening could also be considered a void 2.00 x 4.60 = 9.20m² and because it is > 5m² it would then be deducted from the total of six bays.

Formwork to 1st floor downstand beam

External perimeter at nib

	2/18400	36800 ✓
	2/10400	20800 ✓
	4/2/150	1200 ✓
		59400 ✓
Girth	500	
	400	
	200	
	150	
	300	
	250	
	1800 ✓	

59.40	✓	Formwork, sides and soffits of attached beams, propping, > 3.00m but ne 4.50m high, profiled, four edges, girth 1800, as Detail X [11.18.2.1
1.80	✓	
	106.92 ✓	

Add the sides and soffits first to calculate the total girth of the beam.

The fair finish to the exposed face is measured as an extra over item to allow for the additional cost of this item.

The curved work to the top of the nib is a high labour item and is measured separately to enable a separate price to be calculated.

59.40	✓	Extra over formwork to edge, Detail X, forming curved top to nib, 50mm diameter [11.18.2.1
	59.40 ✓	

59.40	59.40 ✓

Extra over formwork for fair faced finish to exposed face, 300 high
[11.18.2.1

12

Table 8.3 Practical application: Reinforced concrete frame (*Continued*)

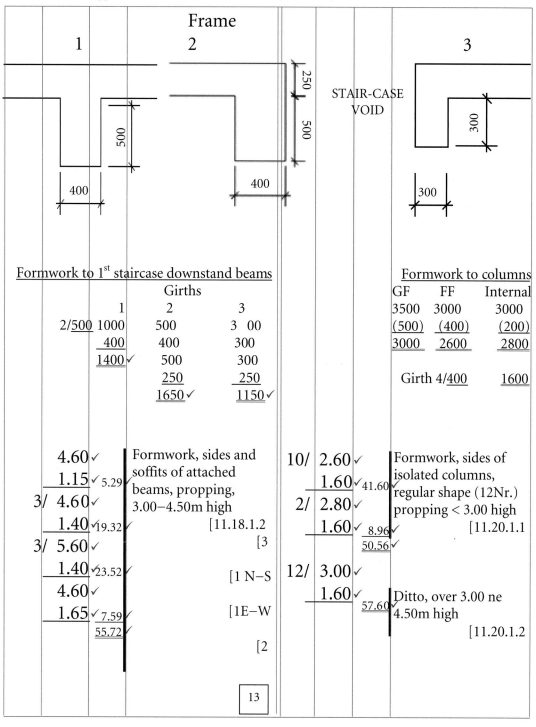

	Frame		
1	2		3

Frame

1   2   3

STAIR-CASE
VOID

Formwork to 1st staircase downstand beams

Girths

	1	2	3
2/500	1000	500	3 00
	400	400	300
	1400 ✓	500	300
		250	250
		1650 ✓	1150 ✓

Formwork to columns

GF	FF	Internal
3500	3000	3000
(500)	(400)	(200)
3000	2600	2800
Girth 4/400		1600

4.60 ✓	
1.15 ✓ 5.29	
3/ 4.60 ✓	
1.40 ✓ 19.32 ✓	
3/ 5.60 ✓	
1.40 ✓ 23.52 ✓	
4.60 ✓	
1.65 ✓ 7.59 ✓	
55.72 ✓	

Formwork, sides and
soffits of attached
beams, propping,
3.00–4.50m high
      [11.18.1.2
      [3

      [1 N–S

      [1E–W

      [2

10/	2.60 ✓
	1.60 ✓ 41.60
2/	2.80 ✓
	1.60 ✓ 8.96 ✓
	50.56 ✓
12/	3.00 ✓
	1.60 ✓ 57.60

Formwork, sides of
isolated columns,
regular shape (12Nr.)
propping < 3.00 high
      [11.20.1.1

Ditto, over 3.00 ne
4.50m high
      [11.20.1.2

13

Table 8.3 Practical application: Reinforced concrete frame (*Continued*)

Frame

All provisional

4✓		Formwork, holes, 50 x 50, horizontal, slab 200mm thick, propping < 3m high [11.31.1				*The formwork necessary to form service holes is included in this section as provisional quantities under item 11.31.* *Section 41 BWIC does not include service holes.*
4✓		Ditto, 400mm thick, do		4✓		Formwork, holes, 50 x 50, vertical, beam 200mm wide, propping < 3m high [11.31.1
4✓		Ditto, 250mm thick, propping 3–4.5m high				
4✓		Formwork, holes, 100 x 100, horizontal, slab 200mm thick, propping < 3m high [11.31.1		4✓		Ditto, 400mm thick, do
				4✓		Ditto, 250mm thick, propping 3–4.5m high
4✓		Ditto, 400mm thick, do				
4✓		Ditto, 250mm thick, propping 3–4.5m high				

14

Table 8.3 Practical application: Reinforced concrete frame (*Continued*)

Frame

		Reinforcement: All				Reinforcement: All Provisional

<table>
<tr><td colspan="2"></td><td colspan="2" align="center">Reinforcement: All
Provisional</td><td colspan="2" align="center">Reinforcement: All Provisional
RC volumes</td></tr>
<tr>
<td><u>25.00</u></td><td>15.00 ✓</td><td colspan="2">Bars, BS 4449, Grade
500C hot rolled, mild
steel, straight, 12mm
[11.33.1.1</td>
<td colspan="2">Roof slab 65m^3
Roof perimeter beams >300 9m^3
Ditto < 300 5m^3
First floor slab 46m^3
Beams 22m^3
Columns <u>11m^3</u>
<u>158m^3</u></td>
</tr>
<tr>
<td><u>2.00</u></td><td>2.00 ✓</td><td colspan="2">Ditto, bent
[11.33.1.2</td>
<td colspan="2">x 2%
= 5m^3 x 7865kg/m^3
= <u>39t</u></td>
</tr>
<tr>
<td><u>2.00</u></td><td>2.00 ✓</td><td colspan="2">Ditto, links
[11.32.1.4</td>
<td colspan="2"></td>
</tr>
<tr>
<td><u>5.00</u></td><td>15.00 ✓</td><td colspan="2">Ditto, BS EN 1.4301
high yield stainless
steel, straight, 12mm
[11.33.1.1</td>
<td colspan="2"><u>Check</u>
158m^3 x 250kg/m^3
=<u>39t</u></td>
</tr>
<tr>
<td><u>2.00</u></td><td>2.00 ✓</td><td colspan="2">Ditto, bent
[11.34.1.2</td>
<td colspan="2">MS 25
2</td>
</tr>
<tr>
<td><u>2.00</u></td><td>2.00 ✓</td><td colspan="2">Ditto, links
[11.34.1.4</td>
<td colspan="2">2
SS 5
2</td>
</tr>
<tr>
<td>18.40
<u>10.40</u></td><td>191.36 ✓</td><td colspan="2">Fabric, BS4483, D98,
1.54kg/m^2, 150 laps
[11.37.1
Ditto, A142,
2.22kg/m^2, do</td>
<td colspan="2">2
<u>2</u>
<u>39</u></td>
</tr>
<tr>
<td>19.00
<u>11.00</u></td><td>209.00 ✓</td><td colspan="2">[11.37.1</td>
<td colspan="2">*The provisional quantities correspond
to the average reinforcement required
for this kind of work*</td>
</tr>
</table>

Both columns of the double dimension paper should be used. In this example the right hand column has often been used for additional notes, diagrams and clarification in several areas.

End of RC frame

15

8.4   SELF-ASSESSMENT EXERCISE: FORMWORK

Measure the formwork to the perimeter of the roof using the plan and section on Drawing SDCO/2/E8/1 in Appendix 8. Please prepare your own measurement using blank double dimension paper in Appendix 1a and also prepare a query sheet of problems that you have encountered. Compare your own work with the proposed solution included in Appendix 8 (Table E8.1) and self-assess your work on the assessment sheet in Appendix 1b.

To provide further assistance there are dedicated websites at http://ostrowski quantities.com and at Wiley Blackwell (http://www.wiley.com/go/ostrowski/measurement). It is hoped that the provision of this will go some way towards explaining the concepts and principles more clearly than using the printed word alone.

9 Brickwork

9.1 Measurement information
- Drawings
- Specification
- Query sheet

9.2 Technology

9.3 Practical application: Brickwork

9.4 Self-assessment exercise: Structural openings in brickwork

9.1 MEASUREMENT INFORMATION

Drawings

See Drawings SDCO/2/9/1 (brickwork plan) and SDCO/2/9/2 (brickwork elevations).

Measurement Using the New Rules of Measurement, First Edition. Sean D.C. Ostrowski.
© 2013 John Wiley & Sons, Ltd. Published 2013 by John Wiley & Sons, Ltd.

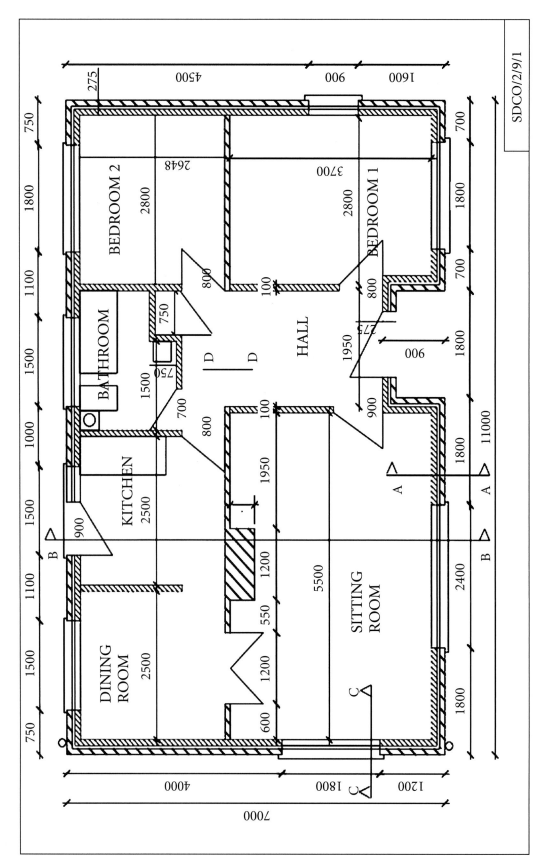

Drawing SDCO/2/9/1 Brickwork plan.

SDCO/2/9/1

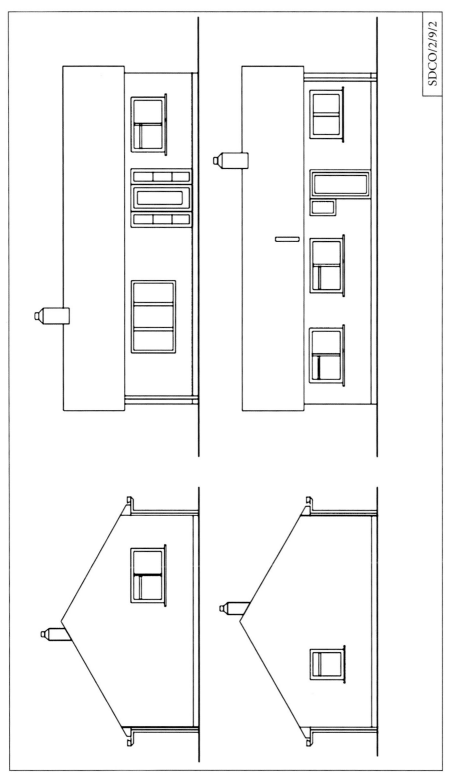

SDCO/2/9/2

Drawing SDCO/2/9/2 Brickwork elevations.

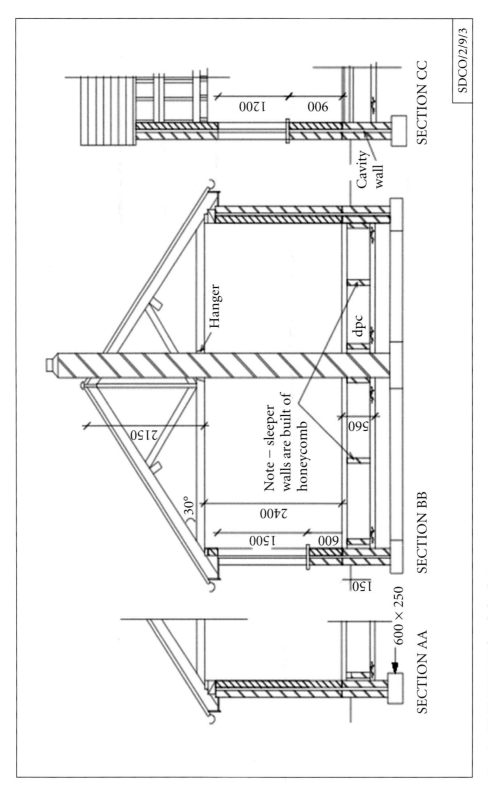

Drawing SDCO/2/9/3 Brickwork details 1.

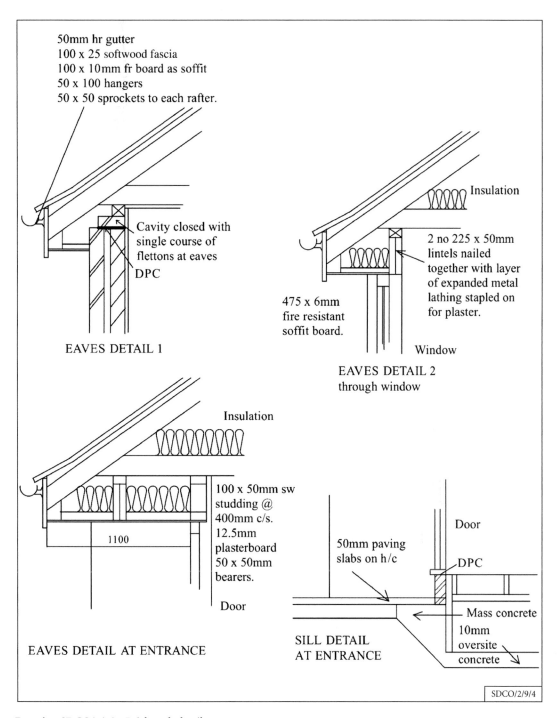

50mm hr gutter
100 x 25 softwood fascia
100 x 10mm fr board as soffit
50 x 100 hangers
50 x 50 sprockets to each rafter.

Cavity closed with
single course of
flettons at eaves

DPC

EAVES DETAIL 1

Insulation

2 no 225 x 50mm
lintels nailed
together with layer
of expanded metal
lathing stapled on
for plaster.

475 x 6mm
fire resistant
soffit board.

Window

EAVES DETAIL 2
through window

Insulation

100 x 50mm sw
studding @
400mm c/s.
12.5mm
plasterboard
50 x 50mm
bearers.

1100

Door

EAVES DETAIL AT ENTRANCE

50mm paving
slabs on h/c

Door

DPC

Mass concrete
10mm
oversite
concrete

SILL DETAIL
AT ENTRANCE

SDCO/2/9/4

Drawing SDCO/2/9/4 Brickwork details 2.

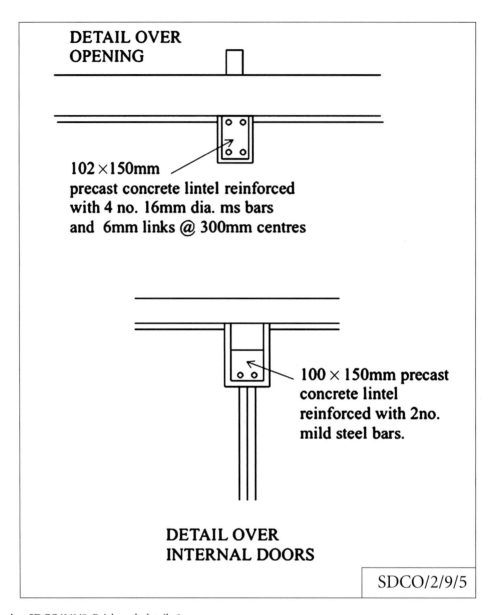

DETAIL OVER OPENING

102×150mm
precast concrete lintel reinforced
with 4 no. 16mm dia. ms bars
and 6mm links @ 300mm centres

100×150mm precast
concrete lintel
reinforced with 2no.
mild steel bars.

DETAIL OVER INTERNAL DOORS

SDCO/2/9/5

Drawing SDCO/2/9/5 Brickwork details 3.

Specification

Table 9.1 Brickwork specification.

Specification S9 Brickwork Superstructure
Precast concrete lintel, 20N/mm², 20mm aggregate, reinforced with mild steel bars, prestressed.
Brickwork above ground to be facing bricks, BS 3921, Prime Cost (PC) £500/1000, stretcher bond, with gauged mortar (1:1:6).
Cavity to be 75mm wide formed with polypropylene ties, 3 nr per m².
Rockwool, or equal and approved, fibre glass insulation batts, BS EN 14162, 50mm thick above DPC.
Damp proof course, Hyload, proprietary, or equal and approved, pitch polymer, BS 743, 150mm laps, bedded in cement mortar (1:3).
Structural timber, sawn softwood, BS EN 14081, grade SC3.
Structural steel lintels, Catnic Type CN7 or equal and approved.
Cavity tray, Cavity Trays Ltd., or equal and approved Type X, polypropylene abutment tray, complete with flashings.

Query sheet

Table 9.2 Brickwork query sheet.

QUERY SHEET	
BRICKWORK SUPERSTRUCTURE	
QUERY **(From the QS)**	**ANSWER** **(From the Architect/Engineer)** **(Assumptions are to be confirmed by QS)**
1. Assumed blockwork specification 2. Assumed cavity ties 2. Assumed insulation specification 3. Assumed PCC specification 4. Assumed steel lintel specifications to front and rear doors	Confirmed xx.xx.xx Confirmed xx.xx.xx Confirmed xx.xx.xx Confirmed xx.xx.xx Confirmed xx.xx.xx

9.2   TECHNOLOGY

The building is a bungalow with external walls of cavity brickwork and blockwork. The internal structural wall is brickwork. Internal partitions are blockwork. Each external structural opening for doors and windows requires a lintel above and a damp proof course around the opening. Each internal structural opening for doors requires a lintel. There is a structural opening in the brickwork crosswall across the hallway that will require a lintel. There is no door and no structural opening between the kitchen and the dining room. Work below the DPC is excluded.

Table 9.3 Practical application: External and internal walls in brickwork.

				Superstructure				
				BUNGALOW				
				BRICKWORK				
				Drawings				
				SDCO/2/9/1 Plan				
				SDCO/2/9/2 Elevation				
				SDCO/2/9/3 Sections				
				SDCO/2/9/4 Details 1–4				
				SDCO/2/9/5 Details 5 & 6				
				Specification S9				*The title page includes the trade name, the name of the contract, the full drawing schedule with dated revisions, the dated specification, the name of the measurer and the date.*
				SDCO May 2012				

Table 9.3 Practical application: External and internal walls in brickwork (*Continued*)

Brickwork

These side casts check the various dimensions horizontally and vertically. One column shows the checking of the arithmetic as an example.
Each column adds up to the same external dimension. This means that the dimensions on the drawing are correct. All the dimensions should be checked in this way.

Horizontal/length dimensions

	1800	275	275 ✓	750
	2400	5500	2500 ✓	1500
	1800	100	100 ✓	1100
	1800	1950	2500 ✓	1500
	700	100	100 ✓	1000
	1800	2800	1500 ✓	1500
	700	275	100 ✓	1100
11000	11000	11000	750 ✓	1800
			100 ✓	750
			2800 ✓	11000
			275 ✓	
			11000 ✓	

Vertical/breadth dimensions

7000	275	4500
	2648	900
	102	1600
	3700	7000
	275	
	7000	

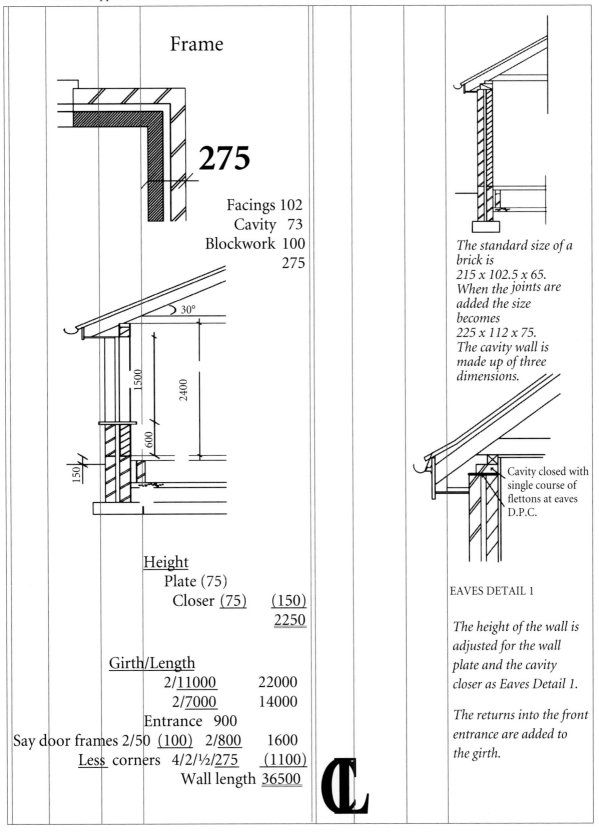

Frame

275

Facings 102
Cavity 73
Blockwork 100
275

30°

1500

2400

600

150

Height
Plate (75)
Closer (75)   (150)
2250

Girth/Length
2/11000   22000
2/7000   14000
Entrance 900
Say door frames 2/50 (100) 2/800   1600
Less corners 4/2/½/275   (1100)
Wall length 36500

The standard size of a brick is
215 x 102.5 x 65.
When the joints are added the size becomes
225 x 112 x 75.
The cavity wall is made up of three dimensions.

Cavity closed with single course of flettons at eaves D.P.C.

EAVES DETAIL 1

The height of the wall is adjusted for the wall plate and the cavity closer as Eaves Detail 1.

The returns into the front entrance are added to the girth.

Table 9.3 Practical application: External and internal walls in brickwork (*Continued*)

			Brickwork				
			Walls, ½b. th, brickwork, skins of hollow walls, Butterfly Brown Brindle bricks, g/m 1:2:9, flush pointed, stretcher bond				*Deductions made for the opening are measured later.*
	36.50						
	2.25		[14.1.1.1				
2/½/	7.00						*On the gable there is no need to deduct for the cavity closer as the external wall is the full height.*
	2.15		[Gable [Band				
2/	7.00						
	.30		2400				
			1500				
			600 2100				
			300				

&

Walls, 100mm th, blockwork, Thermalite Shield blocks, g/m 1:2:9
[14.1.2.1

&

Forming cavity, 73mm wide, formed with polypropylene ties @3/m^2
[14.14.1.1

&

Cavity insulation, Rockwool, 50mm thick, secured to ties
[14.15.1.1

Brickwork

Marking up the drawings checks that all areas have been measured.

Brickwork

DPC/cavity closer

Front & rear elevations   11000
           2/275   (550)
                     10450

Entrance   900
            (275)
            625

2/	10.45	Extra over walls for perimeters, close cavity, 73mm wide, single course 1b wall, commons, g/m 1:2:9, horizontal	The cavity closer dimensions are to the inside face of the cavity wall.
2/	.63		
		[14.12.1.2	The front & rear entrances have no brickwork & therefore no cavity closer or DPC.

Brickwork
Cavity
Blockwork

Gable elevations 7000
         2/275 (550)
            6450

Deductions are taken later.

2/	10.45	DPC < 300mm wide, PVC gauge 1200, bedded in g/m 1:2:9, horizontal	*The DPC has been measured across the gable elevation as well.*
2/	.63		
2/	6.45	[14.16.1.3	

Frame

Pythagoras for the length of the slope at the top of the gable.

Slope of roof

$$x^2 = 2150^2 + [½(7000)]^2$$
$$x^2 = 4.627 + 12.250$$
$$x^2 = 16871$$
$$x = \underline{4108}$$

$^{5.}2/\ \underline{1.20}\ \ \underline{8.40}$

Additional identical dimensions can be added by 'Dotting on' as shown in the example and on the dimensions on subsequent sheets.

$2/2/\ \underline{4.11}$

EO walls for perimeters, close cavity, abd, raking
[14.12.1.2

&

DPC, < 300mm wide, abd, raking
[14.16.1.3

Assumed detail for vertical sides of opening. Confirm on query sheet .

$2/2/\ \underline{2.25}$
$2/\ \underline{1.50}$
$^{2.}3/2/\ \underline{1.20}$
$2/2/\ \underline{1.05}$

EO walls for opening perimeter, close cavity, 75mm wide, blockwork return, vertical
[14.12.1.1

&

DPC, < 300mm wide, abd, vertical
[14.16.1.3

Size of openings		
SR1	2400	1500
SR2	1800	1200
BR1	1800	1200
	900	1200
BR2	1800	1200
DR1	1500	1200
K Wdw	600	1050
K Door	900	2250
Bath	1500	1050
Front	1800	2250

Windows check 8nr✔
Doors check 2nr✔

Table 9.3 Practical application: External and internal walls in brickwork (*Continued*)

<table>
<tr><td colspan="4" style="text-align:center">

Brickwork

Deductions for voids in accordance with NRM Item 11.7.
Voids > 0.50m². NRM 2 p.165 Note 5
</td><td></td></tr>
<tr>
<td valign="top">

Ddt
2.40
1.50
3/1.80
1.20
1.50
1.20
.60
1.05
.90
2.25
1.50
1.05
.90
1.20
1.80
2.25
</td>
<td valign="top">

Ddt
Bkwk walls, ½b, abd
       [14.1.1.1
[SR1

[SR2/BR1/2

[DR1

[Kitchen window. Say

[K door

[Bathroom Say

[BR1

[Front door

    &c

Ddt
Blwk walls, 100mm,
abd
      [14.1.2.1

    &c

Ddt
Cavity, 73 mm, abd
      [14.14.1.1
</td>
<td valign="top">

Ddt
2.40
1·3/1.80
1·2/1.50
.90
</td>
<td valign="top">

Ddt
EO close cavity,
single course 1b
wall, abd, 73mm
wide
      [14.12.1.3

The labour of closing
the cavity over the
structural openings is
not required.

Fair returns, the
additional labour
required to turn the
facework into the door
or window opening is
not a measureable
item.

SR = sitting room
BR = Bedroom
K = Kitchen
</td>
</tr>
</table>

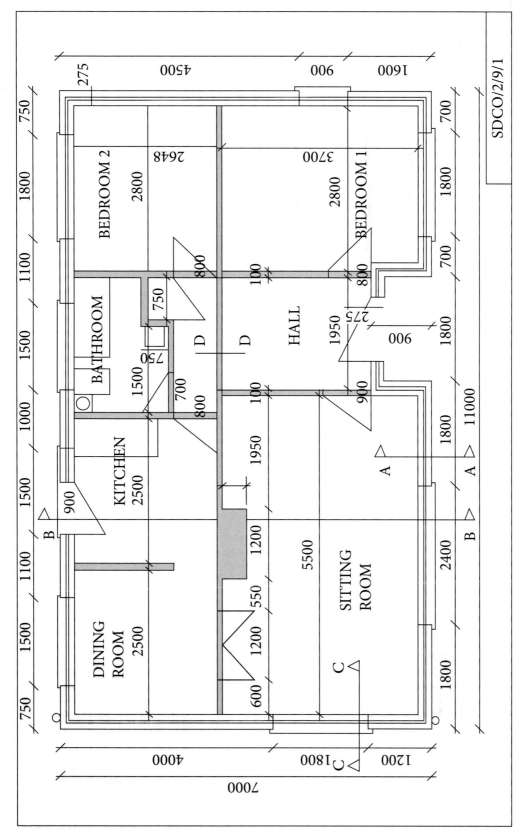

Diagram 9.1 Brickwork internal walls.

SDCO/2/9/1

Table 9.3 Practical application: External and internal walls in brickwork (*Continued*)

Brickwork

Internal walls

Length	5500	
	100	
	100	
	2800	
	8500	

Height	2400	
Less Plate	(75)	
	2325	

		3700
Entrance 900		
	(275)	(625)
		3075

Bath	1500	
	100	
	750	
Linen Say	750	
	750	
	3850	

		Description		
	8.50	Walls, brickwork, ½b thick, vertical, commons, g/m 1:2:9, stretcher bond [14.1.1	3/2.65	Walls, blockwork, 100mm th, vertical, Thermalite Shield insulating blocks, g/m 1:2:9 [14.2.1
Ddt	2.33		2.40	
1.20			2/3.08	
2.00		[L12 door	2.40	
			3.85	
			Ddt 2.40	
			4/ .80	
			2.00	
			.70	
			2.00	
			.90	
			2.00	
			.75	
			2.00	

The work to the chimney is measured as a separate section at the end of this chapter.

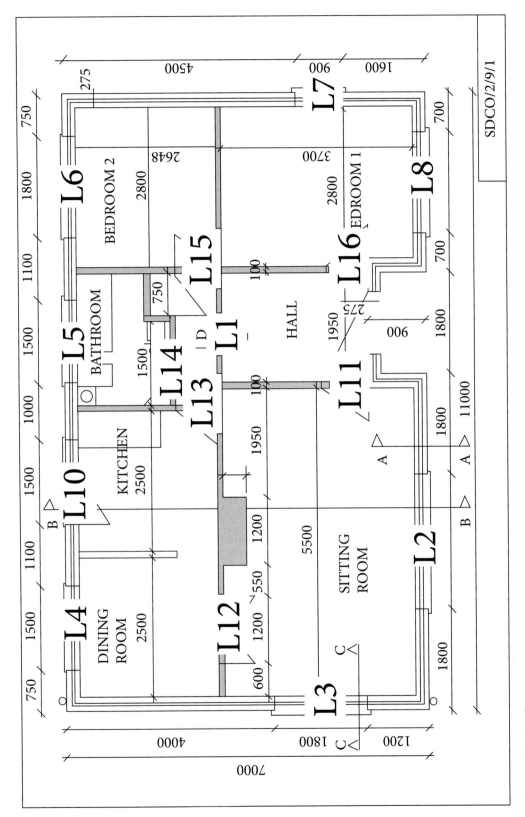

Diagram 9.2 Lintels.

SDCO/2/9/1

Brickwork

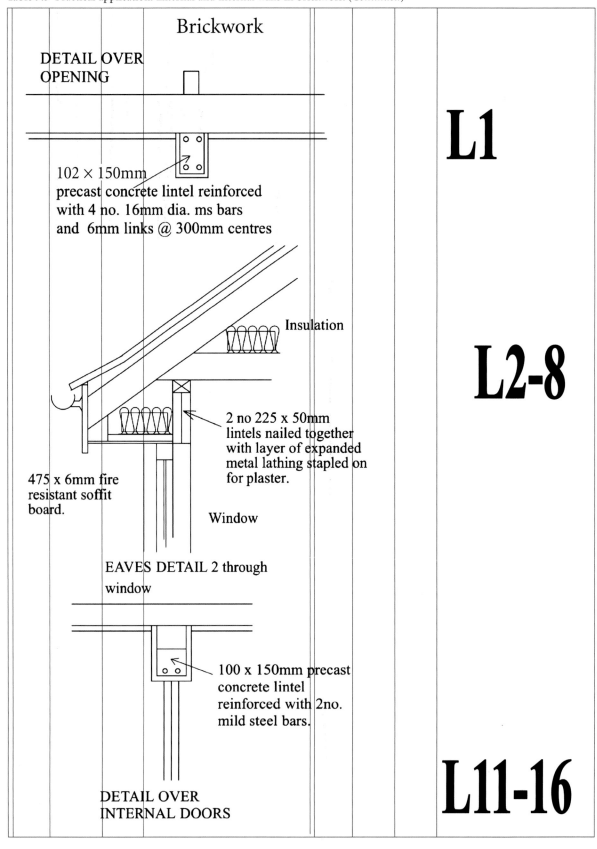

DETAIL OVER OPENING

102 × 150mm precast concrete lintel reinforced with 4 no. 16mm dia. ms bars and 6mm links @ 300mm centres

L1

Insulation

2 no 225 x 50mm lintels nailed together with layer of expanded metal lathing stapled on for plaster.

475 x 6mm fire resistant soffit board.

Window

L2-8

EAVES DETAIL 2 through window

100 x 150mm precast concrete lintel reinforced with 2no. mild steel bars.

DETAIL OVER INTERNAL DOORS

L11-16

Table 9.3 Practical application: External and internal walls in brickwork (*Continued*)

Brickwork

The length of the lintel is the width of the structural opening plus the seating at each end.

Lintel Schedule				Lintel Schedule		
L1	PCC	Hallway		L1	PCC	$1950 + 2/150 = 2250$
L2	Timber	SR1		L2	Timber	$2450 + 2/150 = 2750$
L3		SR2		L3		
L4		DR1		L4		$1500 + 2/150 = 1800$
L5		B1		L5		$1500 + 2/150 = 1800$
L6		BR2		L6		$1800 + 2/150 = 2100$
L7		BR1		L7		$900 + 2/100 = 1100$
L8		BR1		L8		$1800 + 2/150 = 2100$
L9	Steel	Front door		L9	Steel	$1950 + 2/150 = 2250$
L10	Steel	Rear door		L10		$1500 + 2/150 = 1800$
L11	PCC	Hall		L11	PCC	$900 + 2/100 = 1100$
L12		SR		L12		$1200 + 2/150 = 1500$
L13		K		L13		$800 + 2/100 = 1000$
L14		B		L14		$700 + 2/100 = 900$
L15		BR2		L15		$800 + 2/100 = 1000$
L16		BR1		L16		$800 + 2/100 = 1000$

Table 9.3 Practical application: External and internal walls in brickwork (*Continued*)

Brickwork

1	PCC goods, lintel, 102 x 150 x 2250, (PCC 1:2:4, 20mm aggregate), reinforced with 4 nr mild steel bars [L1 [13.1.1.1				Structural timber, lintel, 225 x 50 x 2 nr, spiked, sawn softwood, grade SC3 [16.1.1
			1.10		1100 long [L7
		2/1.80			1800 long [L4, 5
		3/2.10			2100 long [L3,6,8
	Ditto, 100 x 150, do, reinforced with 2 nr mild steel bars [13.1.1.1		2.75		2750 long [L2
1	950 long [L16				
3/ 1	1000 long [L13, 15, 16			1	Isolated metal members, plain metal member, lintel, Catnic CN7, 1800 span [L10 [26.1.1
1	1100 long [L11				
1	1500 long [L12			1	Ditto, 2250 long [L9
—	Lintel check 16 nr ✓				

Brickwork

Chimney

	Projection	450
		(103)
		347

2.40	Chimney stacks, 347 x 1200, vertical, commons, 1:1:6, English bond [14.4.1.1

CHIMNEY FLUE DETAIL

Chimney in roof space

Width reduced from 1200 to 450

Height Say 24 courses x 75 = 1800

5.00	Flue lining, 300 x 300, vitreous clay, rebated, BS 1181, built in & bedded in cm 1:3 [Scaled [14.9.1.1

1.80	Ditto, 450 x 450, abd, g/m 1:1:6, English bond [14.4.1.1

1	Chimney pot, vitreous clay, square base 215 x 215 x 150 diameter, 300 high, bedding and flaunching in c/m 1:3 [14.25.1.1

Exposed chimney stack

	Scaled	2600
Less roof space	(1800)
		800

.80	Ditto, 450 x 450, Butterley Brown Brindle bricks, g/m 1:1:6, flush pointed, English bond [14.4.1.1

Item	Provisional Sum, defined, £5,000 for brickwork fireplace as Sketch nr. Xxx dated xx.xx.xx [14.25.1.1

End of brickwork

9.4   SELF-ASSESSMENT EXERCISE: STRUCTURAL OPENINGS IN BRICKWORK

Measure the deductions for the openings to the brickwork and the labours surrounding them using the plans and sections on Drawings SDCO/2/E9/1, SDCO/2/E9/2, SDCO/2/E9/3 and SDCO/2/E9/4 in Appendix 9/10. Please prepare your own measurement using blank double dimension paper in Appendix 1a and also prepare a query sheet of problems that you have encountered. Compare your own work with the proposed solution in Appendix 9/10 (Table E9.1) and self-assess your work on the assessment sheet in Appendix 1b.

To provide further assistance there are dedicated websites at http://ostrowski quantities.com and at Wiley Blackwell (http://www.wiley.com/go/ostrowski/ measurement). It is hoped that the provision of this will go some way towards explaining the concepts and principles more clearly than using the printed word alone.

10 Openings, Doors and Windows

Measurement Using the New Rules of Measurement, First Edition. Sean D.C. Ostrowski.
© 2013 John Wiley & Sons, Ltd. Published 2013 by John Wiley & Sons, Ltd.

10.1 MEASUREMENT INFORMATION

Drawings

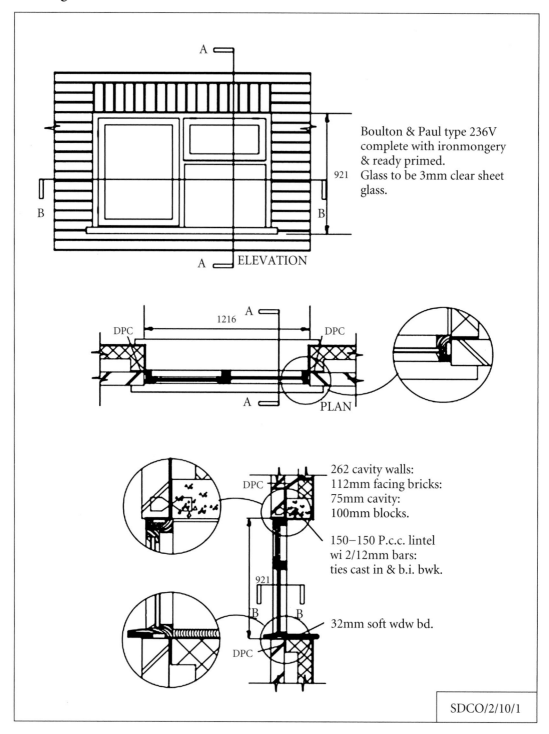

Boulton & Paul type 236V complete with ironmongery & ready primed.
Glass to be 3mm clear sheet glass.

ELEVATION

PLAN

262 cavity walls:
112mm facing bricks:
75mm cavity:
100mm blocks.

150–150 P.c.c. lintel wi 2/12mm bars: ties cast in & b.i. bwk.

32mm soft wdw bd.

SDCO/2/10/1

Drawing SDCO/2/10/1 Windows elevation, section and details.

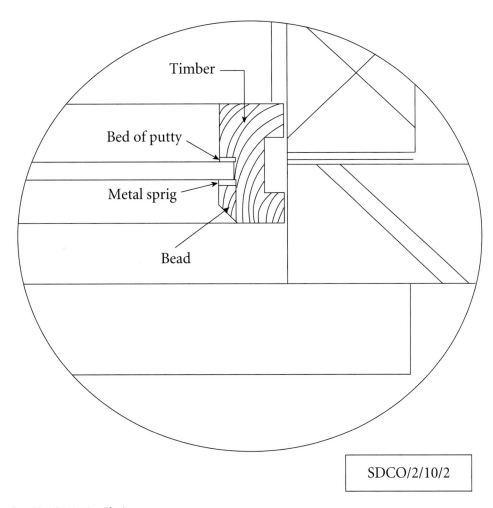

Timber

Bed of putty

Metal sprig

Bead

SDCO/2/10/2

Drawing SDCO/2/10/2 Glazing.

Specification

Table 10.1 Openings and windows specification.

Specification S10 Brickwork and windows
Lintel to be concrete designated mix RC 20/25, 20N/mm², 25mm aggregate with 2 nr 12mm diameter reinforcing bars.
Brickwork outer skin of cavity wall above ground to be facing bricks, Butterley Brown Brindle, BS 3921, stretcher bond, with gauged mortar (1:1:6).
Cavity to be 75mm wide formed with polypropylene ties, 3 nr per m².
Proprietary, or equal and approved, fibre glass insulation batts 50mm thick above DPC.
Blockwork inner skin of cavity to be Thermalite Shield, 100mm thick, with gauged mortar (1:1:6).
Damp proof course, Hyload, proprietary, or equal and approved, pitch polymer, BS 743, 150mm laps, bedded in cement mortar (1:3).
Cavity tray, Cavity Trays Ltd., or equal and approved Type X, polypropylene abutment tray, complete with flashings.
Window boards, wrot softwood, BS EN 14220.
Softwood windows complete by Boulton & Paul, as described, or equal and approved.
Glazing fixed on bed of putty, stainless steel sprigs, and putty and softwood beads as described.
BASIC PAINTING SPECIFICATION
Internal untreated timber Knot, prime and stop, one undercoat oil based paint, one top coat (Kps & ②)
External untreated timber Knot, prime and stop, two undercoats oil based paint, one top coat (Kps & ③)
Internal treated timber Prime, one undercoat oil based paint, one top coat (P & ②)
External treated timber Prime, two undercoats oil based paint, one top coat (P & ③)
Factory primed treated timber Touch up primer, one undercoat oil based paint, one top coat (TU & ②) External, ditto (TU & ③)

Query sheet

Table 10.2 Openings and windows query sheet.

QUERY SHEET	
OPENINGS AND WINDOWS	
QUERY **(From the QS)**	**ANSWER** **(From the Architect/Engineer)** **(Assumptions are to be confirmed by QS)**
1. Assumed blockwork specification	Confirmed xx.xx.xx
2. Assumed cavity ties	Confirmed xx.xx.xx
3. Assumed insulation specification	Confirmed xx.xx.xx
4. Assumed PCC specification	Confirmed xx.xx.xx

10.2   TECHNOLOGY

The building is a bungalow with external walls of cavity brickwork and blockwork.

Each external structural opening requires a large amount of complex work. The structural integrity is maintained with a lintel, the weatherproofing of the walls requires horizontal and vertical damp proofing, doors and windows require decorative and functional details such as cills, heads and steps around the opening.

Windows and doors have to maintain the structural, weatherproofing and security integrity of the brickwork wall whilst providing access and ventilation to the building. The glazing is bedded on putty, secured with small stainless steel nails called sprigs and then surrounded with putty complete with a smooth bevelled finish. This provides added security on ground floor windows where timber beading can easily be removed.

The lintel is cast in situ and is not precast concrete.

Table 10.3 Practical application: Openings, doors and windows.

			Superstructure BUNGALOW OPENINGS AND WINDOWS Drawings SDCO/2/10/1 Elevation SDCO/2/10/2 Details Specification S10 SDCO May 2012				*The title page includes the trade name, the name of the contract, the full drawing schedule with dated revisions, the dated specification, the name of the measurer and the date.*

Openings, doors & windows

Horizontal/length dimensions

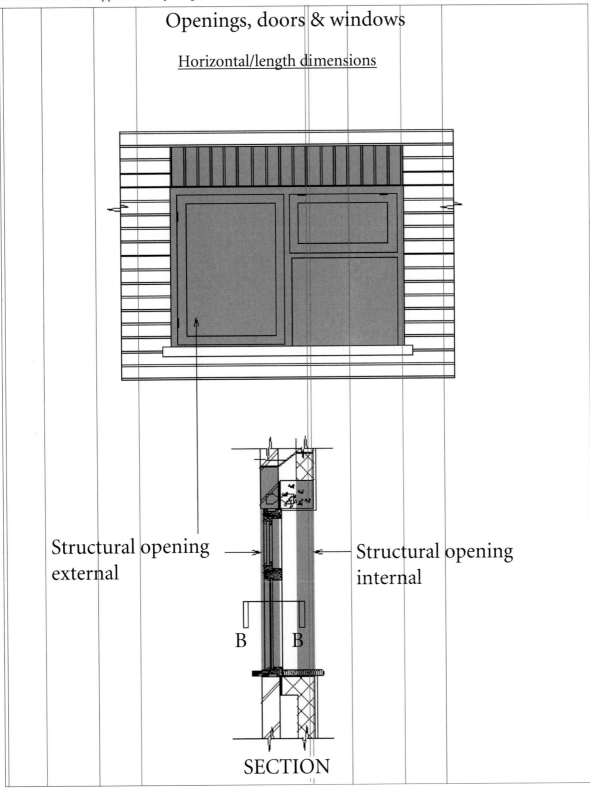

Structural opening external

Structural opening internal

B   B

SECTION

Table 10.3 Practical application: Openings, doors and windows (*Continued*)

		Openings, doors and windows				
		<u>Outer skin brickwork opening</u>				*Brickwork is normally measured over all the openings with deductions taken when the doors and windows are measured. This avoids the additional work of measuring the openings in the brickwork section and measuring the same openings in the windows and doors.*
		Width Height				
		<u>1216</u> 921				
		Header <u>215</u>				
		<u>1136</u>				
<u>Ddt</u> 1.22 <u>1.14</u>		<u>Deduct</u> Walls, ½b. th, bkwk skins of hollow walls, Butterly Brown Brindle bricks, g/m 1:2:9, flush pointed, stretcher bond [14.1.1.1				
	<u>1.22</u>	Bands, header course, 225 high flush pointing, horizontal, entirely of headers [14.7.1.3.2				*Care should be taken to ensure that the brickwork deductions that are measured in this section are not deducted from the brickwork section as well.*

	<u>1.14</u>	Extra over walls for opening perimeters, vertical, closing cavity, with blockwork
		              [14.12.1.2
		
		        &
		
		DPC < 300mm wide, PVC gauge 1200, bedded in g/m 1:2:9, vertical
		              [14.16.1.1
						The protocol used here is that the deductions are shown in the timesing column to ensure that they are not confused with any additional work.

Table 10.3 Practical application: Openings, doors and windows (*Continued*)

Openings, doors and windows

Inner skin brickwork opening

Width		Height
1216		921
	Lintel	150
		1071

Ddt	
1.22	
1.07	

Deduct
Walls, 100th, blwk, vertical, Thermalite Shield, g/m 1:2:9, flush pointed

[14.1.2

&

Forming cavity, 73mm wide, formed with polypropylene ties @3/m^2

[14.14.1.1

&

Cavity insulation, Rockwool, 50mm thick, secured to ties

[14.15.1.1

Table 10.3 Practical application: Openings, doors and windows (*Continued*)

Openings, doors and windows

DPC

262 cavity walls:
112mm facing bricks:
75mm cavity:
100mm blocks.

150 × 150 PCC lintel
wi 2/12mm bars:
ties cast in & b.i. bwk.

	Lintel
	1216
Seating Say 2/ 75	150
	1366

	Cavity tray	
	Blwk	100
	Cavity	50
	Bkwk	112
Lintel (150 + 50)	200
	462

1.37		
.15	Sundry in-situ	
.15	concrete, lintel, < 300	
	thick, horizontal	
	[11.6.1.1	

| 1.37 |
| .46 |

DPC, >300 mm
wide, cavity
gutter, bituminous
felt, horizontal,
dressed (3 nr) into
outer skin of
brickwork and
lintel

[14.16.2.3

3/150   450

1.37	Plain formwork, sides
.45	and soffits of isolated
	beams, 150 x 1
	50, lintel
	[11.17.1

	Window cill
	112
	150
	262

2/ 1.37	Reinforcement, mild
	steel bars, 2 nr, 12mm
	diameter, straight,
	beams, isolated, lintel
	[11.33.1.1

| 1.37 |

DPC, < 300mm
wide, bituminous
felt, horizontal,
dressed (2 nr) into
outer skin of
brickwork and cill

[14.16.1.3

2/1.37 m x_____kg/m
=_____t

Table 10.3 Practical application: Openings, doors and windows (*Continued*)

Openings, doors and windows

1	Windows and frames, softwood, B & P Type 236V, size 1216 x 921 x 50, ironmongery, factory primed, shrink wrapped, plugged and screwed to brickwork, bedding in g/m 1:1:6, mastic pointing to frames, internal and external [23.1.1	*Windows are often glazed and fitted with ironmongery as a prefabricated package. They are measured as NRM 2 item 23.1.1.1.1.1.* *In this example the individual components are measured.*
1.37	Unframed isolated trims, window boards, wrot softwood, 200 x 32, bullnose, tongue & grooved to frame [22.5.1.1	
1	Glass, standard plain glass, 3mm, float, panes 450 x 700, putty, sprigs & beads to timber [27.1.1	*Glazing appears in Section 23.8 and in Section 27. In both cases the panes are enumerated and the pane size is a level 1 description.*
1	Ditto, panes 450 x 175, do	
1	Ditto, panes 550 x 500, do	

Table 10.3 Practical application: Openings, doors and windows (*Continued*)

			Openings, doors and windows					
	1.22 .92		Painting to glazed surfaces, girth > 300mm, internal, softwood, TU & ② [29.2.2.1 & Ditto, external, TU & ③ [29.2.2.2		1		Ironmongery, screwed, B & P SAA Type A, softwood, casement stay & pin [22.22.1 & Ditto, casement fastener, Type B [22.22.1.1
	1.37		Painting, general surfaces, < 300mm softwood, KPS & ② [29.1.1.1				Ditto, casement fastener, Type C [22.22.1.1

End of windows

10.4   SELF-ASSESSMENT EXERCISE: OPENINGS AND DOORS

Measure the deductions for the openings to the brickwork and the labours surrounding them and the hardwood door and frame using the plan and section on Drawing SDCO/2/E9/4 in Appendix 9/10. Please prepare your own measurement using blank double dimension paper in Appendix 1a and also prepare a query sheet of problems that you have encountered. Compare your own work with the proposed solution included in Appendix 9/10 (Table E9.1) and self-assess your work on the assessment sheet in Appendix 1b.

To provide further assistance there are dedicated websites at http://ostrowski quantities.com and at Wiley Blackwell (http://www.wiley.com/go/ostrowski/ measurement). It is hoped that the provision of this will go some way towards explaining the concepts and principles more clearly than using the printed word alone.

11 Flat Roofs

11.1 Measurement information
- Drawings
- Specification
- Query sheet

11.2 Technology

11.3 Practical application: Flat roof

11.4 Self-assessment exercise: Gutters

Measurement Using the New Rules of Measurement, First Edition. Sean D.C. Ostrowski.
© 2013 John Wiley & Sons, Ltd. Published 2013 by John Wiley & Sons, Ltd.

11.1　MEASUREMENT INFORMATION

Drawings

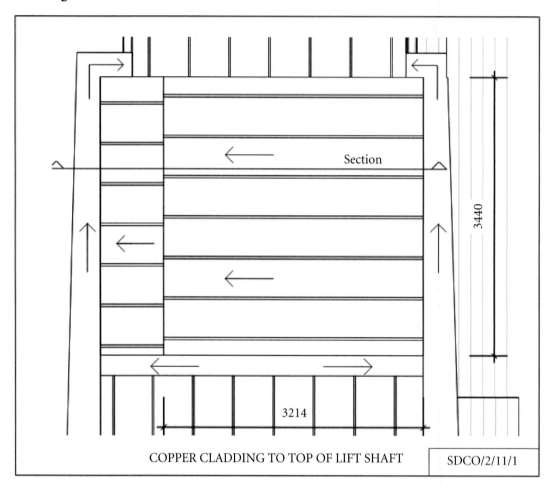

COPPER CLADDING TO TOP OF LIFT SHAFT　　SDCO/2/11/1

Drawing SDCO/2/11/1 Flat roof plan.

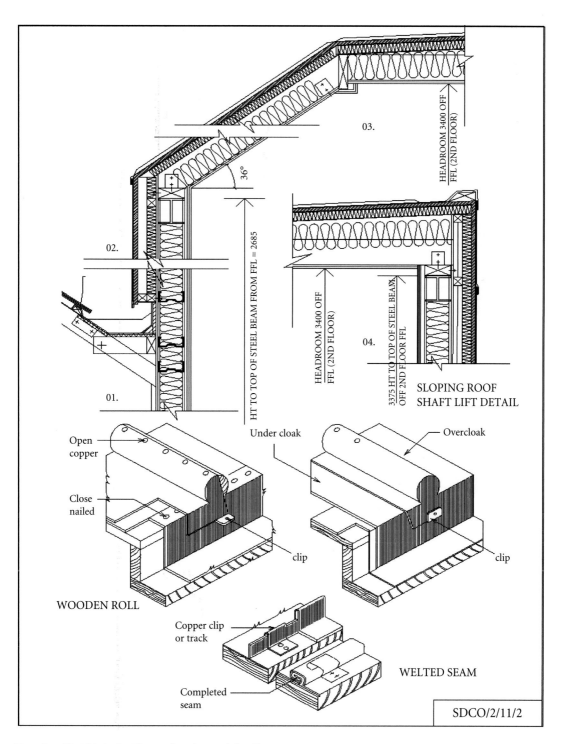

03.

HEADROOM 3400 OFF
FFL (2ND FLOOR)

36°

02.

HT TO TOP OF STEEL BEAM FROM FFL = 2685

HEADROOM 3400 OFF
FFL (2ND FLOOR)

04.

3375 HT TO TOP OF STEEL BEAM
OFF 2ND FLOOR FFL

SLOPING ROOF
SHAFT LIFT DETAIL

01.

Open
copper

Under cloak

Overcloak

Close
nailed

clip

clip

WOODEN ROLL

Copper clip
or track

Completed
seam

WELTED SEAM

SDCO/2/11/2

Drawing SDCO/2/11/2 Flat roof section and details.

Specification

Table 11.1 Flat roof specification.

Specification S11 Copper Roofing
COPPER CLADDING Shingle copper cladding. DIN EN 1172. Pre-patinated finish. Temper of copper: R240: Half-hard. Thickness: 0.7mm. Spacing: 600mm nominal elements centres.
BASE 24mm WBP plywood, BS EN 636. Moisture content: Not more than 22%. Preservative treatment: CCA as section Z12.
UNDERLAY Tyvek® SUPRO breather membrane, or similar approved.
BOTTOM EDGE DETAIL Single welt.
HEAD DETAIL Batten roll as clause 270. Capping: Copper of the same temper, thickness and finish as the roof, in lengths not exceeding 3m, with single lock welt end to end joints.
JOINTS IN CHEEKS Double lock welt joints.
WORKMANSHIP Comply generally with CP 143: Part 12. Fixing of copper: Place two clips to each roof bay. Fold roofing sheets up roll and secure to capping with single lock welt each side. Use solder only where specified. Permitted subject to completion of a 'hot' work permit form and compliance with its requirements. Use clips made from 0.4mm thick austenitic stainless steel or 0.7mm thick copper.
BATTEN ROLL JOINTS: TO COPPER ROOF AREAS Size: 40mm high × 45mm wide tapering to 32mm at apex. Fix to base with brass or stainlesssteel countersunk screws at not more than 500mm centres.
SINGLE LOCK WELT JOINTS: TO HORIZONTAL JOINTS IN CLADDING Form with a 50mm overlap and 25mm underlap. Welt overlap around underlap and lightly dress down to allow freedom of movement.
DOUBLE LOCK WELT JOINTS: TO VERTICAL JOINTS IN CLADDING Form with a 65mm overlap and 40mm underlap. Double welt overlap around underlap and dress down.

Query sheet

Table 11.2 Flat roof query sheet.

QUERY SHEET	
ROOFING	
QUERY **(From the QS)**	**ANSWER** **(From the Architect/Engineer)** **(Assumptions are to be confirmed by QS)**
1. N–S section 2. Junction with guttering 3. Access boarding 4. Vertical height	Assumed xx.xx.11 Included with guttering measured elsewhere Assume nil Assumed 2500. Confirmed

11.2   TECHNOLOGY

Metal sheet fixing remains one of the most complicated items of construction. It uses expensive natural material and requires high levels of workmanship in exposed locations with difficult access using specialist craftsmen.

Table 11.3 Practical application: Flat roof.

			Sheet roof coverings				
			LIFT SHAFT				
			ROOFING				
			Drawings				
			SDCO/2/11/1 Plan				
			SDCO/2/11/2 Section				
			Specification S11				
							The title page includes the trade name, the name of the contract, the full drawing schedule with dated revisions, the dated specification, the name of the measurer and the date.
			SDCO March 2012				

Table 11.3 Practical application: Flat roof (*Continued*) 169

			Roofing				To Take
			Proprietary copper cladding system comprising rolled copper, BS EN 1172, façade grade, R240 half hard temper, 0.7mm thick, pre-patinated finish, underlay Tyvek 'Supro' breather membrane, base 24mm WBP plywood, BS EN 636, not exceeding 22% moisture content, preservative CCA as section Z12, at 600mm centres, fixed in accordance with CP 143: Part 12, horizontal single locked and, vertical double lock welted joints, to roof lift shaft.				Copper sheeting Boarding Battens Rolls Waterproofing Insulation Trims Labours
							This is a proprietary system developed by a specialist subcontractor.
							Care should be taken to ensure the descriptions are in accordance with the architect's design rather than a simple reproduction of the subcontractor's specification information.
	3.21 3.44		Coverings > 500mm wide, horizontal, underlay Tyvek 'Supro', 24mm thick plywood base, pre-patinated finish, to softwood joists at 400mm centres [17.1.1.1.1				

Table 11.3 Practical application: Flat roof (*Continued*)

Roofing

$$\text{Cos } 36° = \frac{800}{h} \text{ (Scaled)}$$

$$h = \frac{800}{0.8090}$$

$$h = \underline{0.988}$$

Ditto, sloping, 36°
pitch, do

[17.2.1.1.1

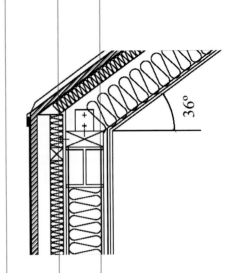

Perimeter vertical
cladding

$$988^2 = 800^2 + x^2$$
$$x^2 = 976144 - 640000$$
$$x = \sqrt{336144}$$
$$x = \underline{579}$$

3214	2500
800	(579)
4014	1921

Ditto, vertical,
50 x 50 softwood
battens at 600mm
centres

[17.1.2.1.1

[N & S elevation

[West

[East

The following items are measurable in NRM 2 and comprise high labour content.

They incorporate the standard details shown in the drawings.

The provisional items are measured to provide measurable items that may be necessary and rates for work that may be included in the variations section of the final account.

Table 11.3 Practical application: Flat roof (*Continued*)

Sheet roof coverings

2/ 3.21 3/ 3.44	Ditto, extra over for forming single welted drip, 35mm, do [17.3.1.1	7/3.21	Ditto, rolls, wood core, horizontal softwood, preservative, 40 x 45 tapering to 32 at apex, fixed with brass screws ne 500mm centres, do [17.3.1.1.1	
2/ 3.44	Ditto, single welted joints horizontal, 50mm overlap, 25mm underlap, do (PROVISIONAL) [17.3.1.1.1	7/1.00	Ditto, sloping, do	
		4/7/2.30	Ditto vertical, do	
4/ 2.50	Ditto, double welted joints to vertical face, 65mm overlap, 40mm underlap, do (PROVISIONAL) [17.3.1.1.4	3.44	Ditto, boundary work, copper cap flashing, horizontal, 300mm girth, 3x dressing, single welted end, fixed with copper straps, 150 x 50 softwood profiled upstand [17.4.1.1.1	
		3.44	Ditto, leading edge to edge of slope, 144°, do	
		3.44 2/3.21	Ditto, double welted joint, 90°, do	

End of flat roof

11.4   SELF-ASSESSMENT EXERCISE: GUTTERS

Measure the guttering using the plan and section on Drawings SDCO/2/11/1 and SDCO/2/11/2. Please prepare your own measurement using blank double dimension paper in Appendix 1a and also prepare a query sheet of problems that you have encountered. Compare your own work with the proposed solution in Appendix 11 (Table E11.1) and self-assess your work on the assessment sheet in Appendix 1b.

   To provide further assistance there are dedicated websites at http://ostrowski quantities.com and at Wiley Blackwell (http://www.wiley.com/go/ostrowski/ measurement). It is hoped that the provision of this will go some way towards explaining the concepts and principles more clearly than using the printed word alone.

12 Pitched Roofs

Measurement Using the New Rules of Measurement, First Edition. Sean D.C. Ostrowski.
© 2013 John Wiley & Sons, Ltd. Published 2013 by John Wiley & Sons, Ltd.

12.1 MEASUREMENT INFORMATION

Drawings

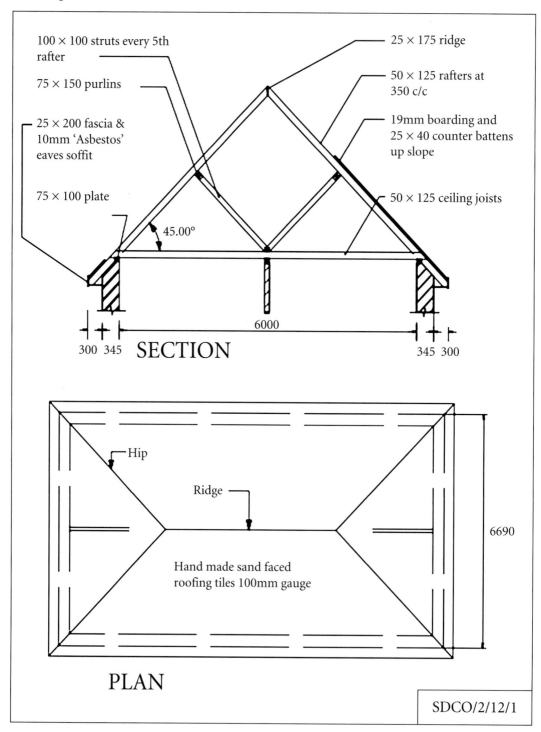

100 × 100 struts every 5th rafter

75 × 150 purlins

25 × 200 fascia & 10mm 'Asbestos' eaves soffit

75 × 100 plate

25 × 175 ridge

50 × 125 rafters at 350 c/c

19mm boarding and 25 × 40 counter battens up slope

50 × 125 ceiling joists

45.00°

6000

300 345   SECTION   345 300

Hip

Ridge

Hand made sand faced roofing tiles 100mm gauge

6690

PLAN

SDCO/2/12/1

Drawing SDCO/2/12/1 Pitched roof plan and section.

Specification

Table 12.1 Pitched roof specification.

Specification S12 Pitched Roofing
Structural timber, sawn softwood, BS EN 14081 grade SC3.
Hanson clay roof tiles type plain, 'Dark Heather' as described or equal and approved.
Underlay, PVC Type 1F, reinforced.
Boarding to be plywood, external quality.
All timber to be treated with preservative at the factory.
Asbestos free strip undercloak, dry soffit board system.

Query sheet

Table 12.2 Pitched roof query sheet.

QUERY SHEET	
ROOFING	
QUERY (From the QS)	**ANSWER** (From the Architect/Engineer) (Assumptions are to be confirmed by QS)
1. Assumed timber specification	Confirmed xx.xx.11
2. Assumed boarding specification	Confirmed xx.xx.xx
3. Hanson clay roof tiles type plain 'Dark Heather' as described or equal and approved	Confirmed xx.xx.xx Confirmed xx.xx.xx
4. Assumed sarking felt specification	Confirmed xx.xx.xx
5. Assumed hipped iron specification	Confirmed xx.xx.xx
6. Assumed soffit board specification	Confirmed xx.xx.xx
7. Assumed painting specification	Confirmed xx.xx.xx
8. Assuming bearing of central plate, 100 mm	Confirmed xx.xx.xx
9. Size of hip timber, assumed 200 × 38	Confirmed xx.xx.xx
10. Height of building for rainwater pipe	Assumed 5400. Confirmed.

12.2   TECHNOLOGY

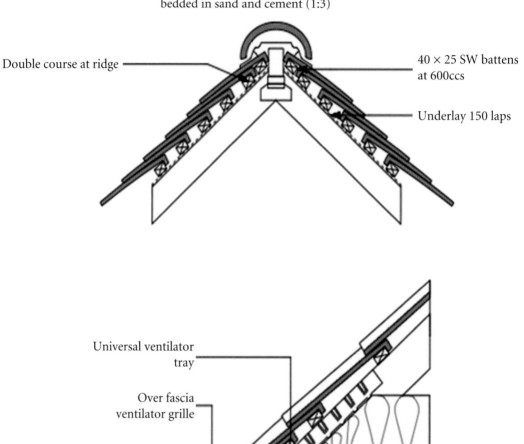

Half round ridge tiled. Mechanical fixing bedded in sand and cement (1:3)

Double course at ridge

40 × 25 SW battens at 600ccs

Underlay 150 laps

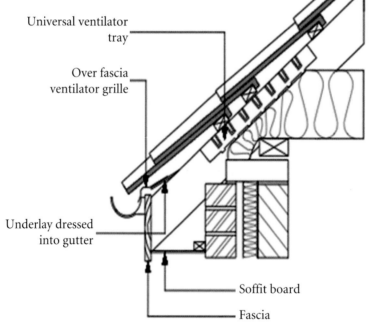

Universal ventilator tray

Over fascia ventilator grille

Underlay dressed into gutter

Soffit board

Fascia

SDCO/2/12/2

Diagram 12.1 Fixing details for a typical sloping roof with clay tiles.

Table 12.3 Practical application: Pitched roof.

			Superstructure					
								The title page includes the trade name, the name of the contract, the full drawing schedule with dated revisions, the dated specification, the name of the measurer and the date.
			BUNGALOW					
								A 'to take' list acts as a check list.
			ROOFING					
			Drawings					To Take
			SDCO/2/12/1 Plan and section					Timber
								Plates
			Specification S12					Rafters
								Purlins
								Binders
								Struts
								Ceiling joists
								Ridges
								Hips
								Coverings
								Boarding
								Battens /counterbattens
			SDCO					Sarking felt
			May 2012					Tiles
								Roof iron
								Painting
								Guttering & pipework
								Gutters
								Pipework
								Bargeboards
								Gutter boards
								Chimneys

Table 12.3 Practical application: Pitched roof (*Continued*)

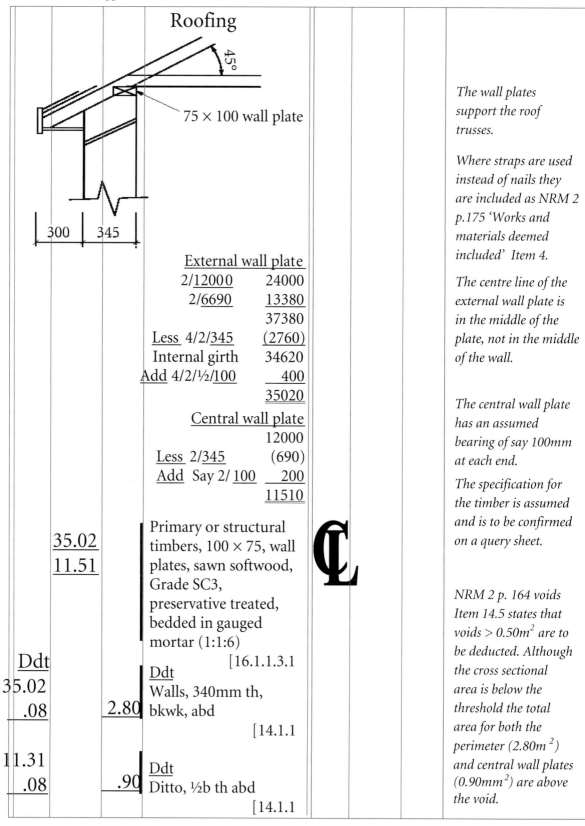

Roofing

45°

75 × 100 wall plate

| 300 | 345 |

External wall plate

2/12000	24000
2/6690	13380
	37380
Less 4/2/345	(2760)
Internal girth	34620
Add 4/2/½/100	400
	35020

Central wall plate

	12000
Less 2/345	(690)
Add Say 2/100	200
	11510

$\underline{35.02}$
$\underline{11.51}$

Primary or structural timbers, 100 × 75, wall plates, sawn softwood, Grade SC3, preservative treated, bedded in gauged mortar (1:1:6)

[16.1.1.3.1

Ddt
$\underline{35.02}$
.08

Ddt
Walls, 340mm th, bkwk, abd

2.80

[14.1.1

11.31
.08

Ddt
Ditto, ½b th abd

.90

[14.1.1

ℭℒ

The wall plates support the roof trusses.

Where straps are used instead of nails they are included as NRM 2 p.175 'Works and materials deemed included' Item 4.

The centre line of the external wall plate is in the middle of the plate, not in the middle of the wall.

The central wall plate has an assumed bearing of say 100mm at each end.

The specification for the timber is assumed and is to be confirmed on a query sheet.

NRM 2 p. 164 voids Item 14.5 states that voids > 0.50m^2 are to be deducted. Although the cross sectional area is below the threshold the total area for both the perimeter (2.80m^2) and central wall plates (0.90mm^2) are above the void.

Table 12.3 Practical application: Pitched roof (*Continued*)

Roofing

The lengths of the rafters in the ends of a hipped roof are the same as if the ridge continued to the end to form a gable end. The rafter numbers combine together as demonstrated in the lower sketch. Trigonometry provides the calculation for the length of the sloping rafter.

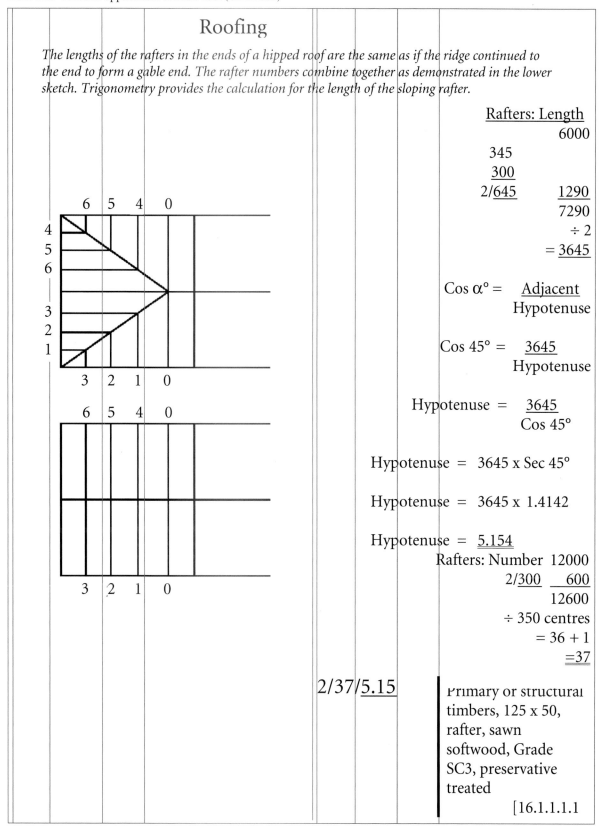

Rafters: Length

$$6000$$
$$345$$
$$\underline{300}$$
$$2/\underline{645} \qquad 1290$$
$$7290$$
$$\div 2$$
$$= \underline{3645}$$

$$\text{Cos } \alpha° = \frac{\text{Adjacent}}{\text{Hypotenuse}}$$

$$\text{Cos } 45° = \frac{3645}{\text{Hypotenuse}}$$

$$\text{Hypotenuse} = \frac{3645}{\text{Cos } 45°}$$

$$\text{Hypotenuse} = 3645 \times \text{Sec } 45°$$

$$\text{Hypotenuse} = 3645 \times 1.4142$$

$$\text{Hypotenuse} = \underline{5.154}$$

Rafters: Number   12000
$$2/\underline{300} \qquad \underline{600}$$
$$12600$$
$$\div 350 \text{ centres}$$
$$= 36 + 1$$
$$=\underline{37}$$

2/37/5.15

Primary or structural timbers, 125 x 50, rafter, sawn softwood, Grade SC3, preservative treated

[16.1.1.1.1

Table 12.3 Practical application: Pitched roof (*Continued*)

Roofing

<u>Purlins</u>

		12000			<u>Struts</u>
	<u>Less</u> wall thickness	2/<u>345</u>			$3000^2 = x^2 + x^2$
		(690)			$9 = 2x^2$
		11310			$4500 = x^2$
					<u>2.12</u> = x

$37 \div 5 = \underline{\underline{7}}$

<u>Binder</u>

		12000
<u>Less</u>	2/<u>345</u>	(690)
		11310
Seating 2/<u>100</u>		200
		11510

2/<u>11.31</u>	Primary or structural timbers, 150 x 75, purlin, sawn softwood, Grade SC3, preservative treated		2/7/<u>2.12</u>	Primary or structural timbers, 100 x 100, strutting, sawn softwood, Grade SC3, preservative treated	
<u>11.51</u>	[16.1.1.1.1			[16.1.1.8.1	

<u>Ceiling Joists</u>

			6000
<u>100</u>			
<u>50</u>	2/<u>150</u>	300	
		6300	

<u>Ridge</u>

		12000
2/<u>300</u>		600
		12600
<u>Less</u> Scaled 2/<u>3600</u>		(7200)
		<u>5400</u>

37/ <u>6.30</u>	Ditto, abd, SSW, 125 x 50, joists		<u>5.40</u>	Ditto, abd, SSW, 125 x 50, ridge	
	[16.1.1.4.1			[16.1.1.1.1	

Table 12.3 Practical application: Pitched roof (*Continued*)

Roofing

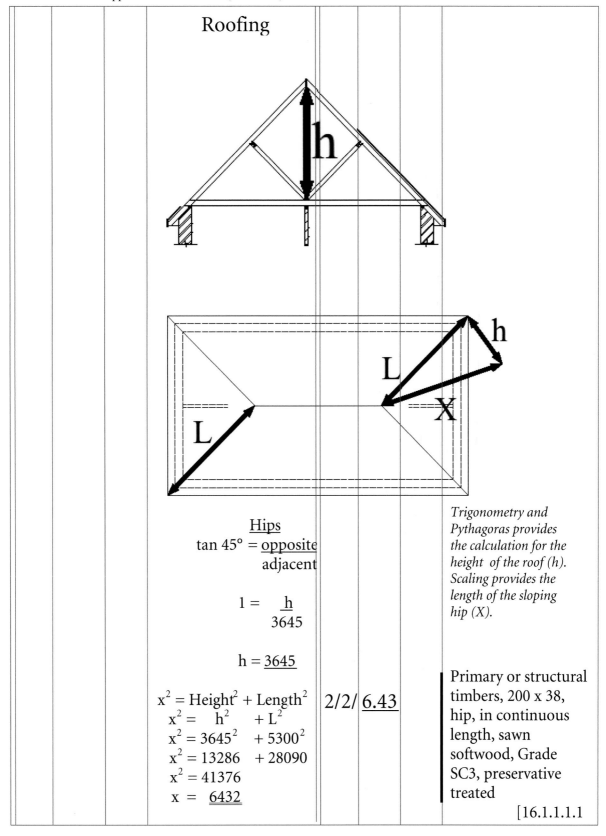

Trigonometry and Pythagoras provides the calculation for the height of the roof (h). Scaling provides the length of the sloping hip (X).

$$\underline{\text{Hips}}$$

$$\tan 45° = \frac{\text{opposite}}{\text{adjacent}}$$

$$1 = \frac{h}{3645}$$

$$h = \underline{3645}$$

$$x^2 = \text{Height}^2 + \text{Length}^2$$
$$x^2 = h^2 + L^2$$
$$x^2 = 3645^2 + 5300^2$$
$$x^2 = 13286 + 28090$$
$$x^2 = 41376$$
$$x = \underline{6432}$$

2/2/ 6.43

Primary or structural timbers, 200 x 38, hip, in continuous length, sawn softwood, Grade SC3, preservative treated

[16.1.1.1.1

Table 12.3 Practical application: Pitched roof (*Continued*)

Roofing

2/12.60 __5.15__	Boarding, >600mm wide, 150 x 19mm thick, sloping, plywood, external quality, preservative, fixed at 350 c/cs with 2 nr galvanised steel nails [16.4.2.2	*Different elements of the roof coverings have special fixings which are to be included in the description.*
	&	
	Hanson plain clay, 'Dark Heather' roof tiling fixed in accordance with the manufacturer's instructions	
5.40	Roof coverings, 45° pitch, clay roof tiles, 254 x 165 plain sand faced roofing tiles with 65mm end lap, tiles nailed every 5th course with 2 nr 38 x 12g aluminium alloy nails to 25 x 40mm battens, sawn softwood, at 265 centres, PVC reinforced underlay, type 1F [18.1.1.1	Hanson plain clay roof tiling abd Ditto, boundary work, ridge tile, 450mm half round, horizontal, with mechanical fixing, bedded & pointed in g/m (1:1:6) [18.3.1.3.1
2/2/ 6.41		Ditto, boundary work, hip tile, third round, fixed abd [18.3.1.3.2

Table 12.3 Practical application: Pitched roof (*Continued*) 183

4/ 1			Roofing		*Metalwork and painting are included in the roof measurement because they will be done at the same time as the roof, because the scaffolding will still be in place to provide access for the work.*

Fittings, galvanised hip iron, 250 x 38 x 8, decorative, fixed with stainless steel screws
[18.4.1.4.2

&

Painting, metalwork, isolated, hip iron, ornamental, P & ③
[29.8.3

38.58

Perimeter
2/12000 24000
6/6690 13380
37380
4/300 1200
38580

Clay roof tiles, abd, boundary work, eaves, double course of tiles, fixed with s/s nails
[18.3.1.2.2

&

Carpentry casings, < 600, fascia board, 200 x 25, horizontal, 1 x grooved, wrot softwood, preservative, abd
[16.4.1.2.1

&

Ditto, asbestos free, soffit board, 300 x 10, fixed with s/s screws
[16.4.1.1.

&

Ditto, batten, 50 x 50, plugged and screwed to brickwork
[16.3.1.2

Table 12.3 Practical application: Pitched roof (*Continued*)

Roofing

		Soffit 300	2/ 5.40	Pipework, 75mm dia., straight, Key Terrain, PVC pipes, to brickwork with brass screws
		Fascia 200		[Assumed
		500		[33.1.1.1.1
	38.58 .50	Painting, gs, > 300 girth, external, timber, KPS & ③ [29.1.2.2		
			2/ 1	Pipework ancillaries, offset, 75 dia x 45° [33.2
	38.58	Gutters, 100mm dia., straight, push fit, Terrain, PVC gutters, half round, to woodwork with brass screws [33.5.1.1.1	2/ 1	Ditto, joint to drainage, c/m 1:3, 100mm dia
			Item	Mark out for RW installation [33.8
4/ 1		Gutter ancillaries, angle [33.6	Item	T & C [33.10
2/ 1		Ditto, outlet		
2/ 1		Ditto, galvanised wire balloon		

End of pitched roof

12.4   SELF-ASSESSMENT EXERCISE: TILING TO ROOF

Measure the roof tiling and gutters using the plan and section on Drawing SDCO/2/E12/1 in Appendix 12. Please prepare your own measurement using blank double dimension paper in Appendix 1a and also prepare a query sheet of problems that you have encountered. Compare your own work with the proposed solution in Appendix 12 (Table E12.1) and self-assess your work on the assessment sheet in Appendix 1b.

To provide further assistance there are dedicated websites at http://ostrowski quantities.com and at Wiley Blackwell (http://www.wiley.com/go/ostrowski/measurement). It is hoped that the provision of this will go some way towards explaining the concepts and principles more clearly than using the printed word alone.

13 Steelwork

13.1 Measurement information
  - Drawings
  - Specification
  - Query sheet
13.2 Technology
13.3 Practical application: Agricultural building
13.4 Self-assessment exercise: Steelwork to pergola

13.1 MEASUREMENT INFORMATION

Drawings

See Drawing SDCO/2/13/1 (steelwork plan and sections).

Measurement Using the New Rules of Measurement, First Edition. Sean D.C. Ostrowski.
© 2013 John Wiley & Sons, Ltd. Published 2013 by John Wiley & Sons, Ltd.

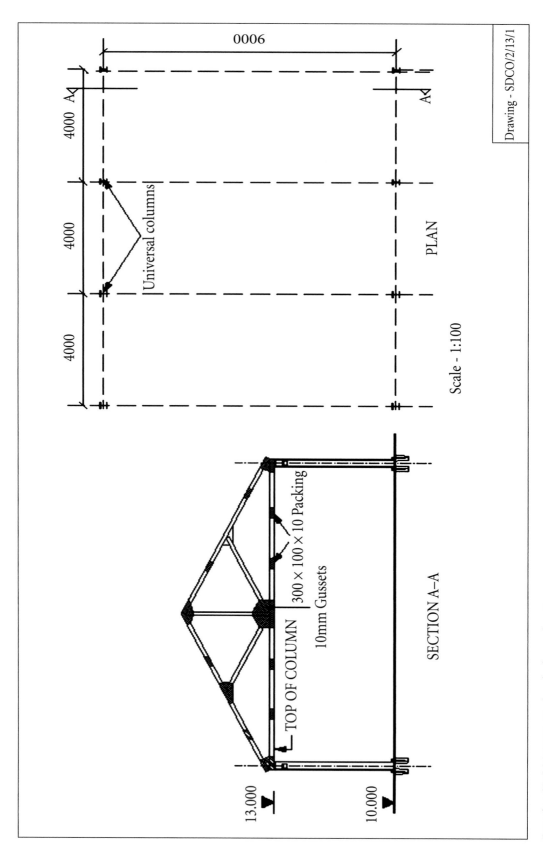

Drawing SDCO/2/13/1 Steelwork plan and sections.

Specification

Table 13.1 Steelwork specification.

Specification S13 Steelwork
All structural steelwork to be BS EN 10025: 2004, Grade 275, hot rolled, welded fabrication, or equal and approved.
Columns 203 × 203 × 86kg/m Universal column.
Beams 203 × 133 × 30kg/m Universal beam.
Cap plate 20mm plate cut to profile.
Cleats 127 × 127 × 23.97kg/m angle.
Base plate 40 × 400 × 400 holed for 4 × M20 bolts.
TRUSS MEMBERS
Tie: 2 Nr. 100 × 100 × 15kg/m angle set back to back.
Rafter: 2 Nr. 100 × 100 × 12kg/m angle set back to back.
Hanger Tie: 2 Nr. 100 × 100 × 15kg/m angle set back to back.
Struts: 2 Nr. 100 × 100 × 15kg/m angle set back to back.
Gussets: 10mm plate cut to profile.
Packings: 10 × 100 × 100mm plate.

Query sheet

Table 13.2 Steelwork query sheet.

QUERY SHEET: STEELWORK	
QUERY (From the QS)	**ANSWER** (Assumptions are to be confirmed)
1. Surface treatment, intumescent paint. 2. Proprietary epoxy resin grout.	Confirmed xx.xx.xx Confirmed xx.xx.xx

13.2 TECHNOLOGY

The building is a simple framework of steel columns and beams. The roof is made up of prefabricated trusses using steel angle fixed back to back. The complete prefabricated truss is erected on site. The surface finish is shot blasting and epoxy zinc primer at the factory and an intumescent painted finish on site.

The columns are cast into the concrete base (measured separately in the Substructure) using a proprietary epoxy resin grout specified by the Engineer. The work can only be done when the concrete base is cast but the measurement, which covers several different trades, is included in the steelwork.

13.3   PRACTICAL APPLICATION: AGRICULTURAL BUILDING

Table 13.3 Practical application: Steelwork to agricultural building.

			Steelwork					
			STEELWORK FRAME					
			Drawings					
			SDCO/1/13/1 Plan					
			SDCO/1/13/2 Section					
			Specification S13					
								The title page includes the trade name, the name of the contract, the full drawing schedule with dated revisions, the dated specification, the name of the measurer and the date.
			SDCO					
			May 2012					

Table 13.3 Practical application: Steelwork to agricultural building (*Continued*)

Structural metalwork

		Structural steel framing of welded construction to BS EN 10025: 2004, Grade 275, hot rolled, welded fabrication.	NRM 3.3.2e indicates tonne are to two decimal points. The steel industry standard uses three decimal points.
		Section A–A 13000 (10000) 3000	
8/ 3.00 24.00		Framed members, permanent erection on site, length 1–9m, weight 50–100kg/m, columns, 203 x 203mm x 86kg/m [15.2.2.3.1	Material prices are based on different cross sectional sizes and weights per linear metres.
			Different sizes and weights have been separated.
		$\dfrac{24.00 \ \text{x} \ 86}{1000} = 2.06\text{t}$	The price for the erection is based on the total weight.
		Ditto, beams, 203 x 133mm x 30kg/m, do [15.2.2.3.2	
6/ 4.00 24.00			
			The calculation of the total weight also provides the allowance for the fittings.
		$\dfrac{24.00 \ \text{x} \ 30}{1000} = .72\text{t}$	

		2.064
		  .720
		 2.784 x 10% = .28t
		 Allowance for fittings, to framed members, 10%
		              [15.5.1.2
			A detailed measurement for the fittings can provide a more accurate price as they are high value items, as follows.

.28			

Structural metalwork

Cap plate ⎯
Angle cleat ⎯

Base plate ⎯

8/	.40		Base plate 40mm thick
	.40	1.28	
6/2/2/	.13	3.12	Angle cleat 127 x 127
8/	.20		Cap plate
	.20	0.37	

The fittings are then high value items and a detailed measurement for them can provide a more accurate price.

1.28 x Say 50 kg/m^2 = 64
3.12 x 23.97 kg/m^2 = 75
.32 x Say 50 kg/m^2 = 16
155 kg

.16		Allowance for fittings, calculated weight, to framed members.
		[15.5.1.1

Structural metalwork

| 8/ | 4 | | Holding down bolts, 20 'Anchor' bolt, 400 long with 100 x 100 x 20 plate, nut and washer
 [15.10.1 | M20 × 400 holding down bolt with 100 × 100 × 20 washer welded on and nut and washer | | | | | |

Epoxy resin grouted bolt boxing

&

Formwork, complex shapes, mortice girth < 50 diameter, depth 250–500
[11.30.1

&

These complex items require measurement from several trades.

Filling hollow sections, in-situ concrete filling, mortices, 400mm deep, epoxy resin based mortar
[11.43.1.1

8

In-situ concrete grouting, stanchion bases, cement/sand, (1:2)
[11.42.1.1

				Structural metalwork				

Height of truss scaled <u>2200</u>

Horizontal length of truss 9000
Size of column 2/½/<u>203</u>   <u>203</u>
              9203
            ÷ 2
          = <u>4602</u>

$x^2 = 4.60^2 \times 2.20^2$
$x^2 = 21.16 \times 4.84^2$
$x^2 = 26.00$
$x = \sqrt{26.00}$
$x = \underline{5.10}$

*'The types and sizes
of all structural
members and their
positions in relation
to each other.'*
[NRM p.169 Cl. 15.2

*This indicates that
the individual
members require
measurement.*

4/2/	<u>9.20</u>	73.60	Framed members, permanent erection on site, in 4 Nr. trusses, 9000 span, 1–10.00m long, < 25kg/m, trusses, 100 x 100 x 15kg/m, equal angle as tie [15.2.2.1.7					<u>73.60</u> x 15kg/m = 1,104

<u>73.60</u> x 15kg/m = 1,104
<u>81.60</u> x 12kg/m =   979
<u>17.60</u> x 15kg/m =   264
<u>43.20</u> x 15kg/m = __648__
            2.995
              ÷
            1000
          = <u>3.00</u>t

4/2/2/	<u>5.10</u>	81.60	Ditto, 100 x 100 x 12kg/m, equal angle as rafter					
4/2/	<u>2.20</u>	17.60	Ditto, 100 x 100 x 15kg/m, equal angle as hanger		3.00			Allowance for fittings, calculated weight, to framed members [15.5.1.1
4/2/2/	<u>2.70</u>	43.20	Ditto, 100 x 100 x 15kg/m angle as strut					

Structural metalwork

Columns	4/209	836
	2/222	444
		1280

Beams	4/ 207	828
	2/134	268
		1096

Trusses	6/ 101	606

Lengths	73.76
	81.60
	17.60
	43.20
	216.16

209

222

101

101

8/	3.00
	1.28
6/	4.00
	1.10
	216.16
	.61

Surface treatment, shot
blasting and priming
to SA2.5, one coat two
pack epoxy zinc
phosphate primer, at
factory
　　　　　[15.15.4.2.1

&

Touch up primer, and
intumescent paint fire
protection, 30 minutes,
spray applied, on site
　　　　　[15.15.2.1

*Both columns of the double dimension paper should be used. In this example the right hand column
has often been used for additional notes, diagrams and clarification in several areas.*

End of structural metalwork

13.4   SELF-ASSESSMENT EXERCISE: STEELWORK TO PERGOLA

Calculate the length and weight of the steelwork to be constructed for a pergola as shown on Drawing SDCO/2/E13/1 in Appendix 13. Please prepare your own measurement using blank double dimension paper in Appendix 1a and also prepare a query sheet of problems that you have encountered. Compare your own work with the proposed solution in Appendix 13 (Table E13.1) and self-assess your work on the assessment sheet in Appendix 1b.

   To provide further assistance there are dedicated websites at http://ostrowski quantities.com and at Wiley Blackwell (http://www.wiley.com/go/ostrowski/ measurement). It is hoped that the provision of this will go some way towards explaining the concepts and principles more clearly than using the printed word alone.

14 Partitions

Measurement Using the New Rules of Measurement, First Edition. Sean D.C. Ostrowski.
© 2013 John Wiley & Sons, Ltd. Published 2013 by John Wiley & Sons, Ltd.

14.1 MEASUREMENT INFORMATION

Drawings

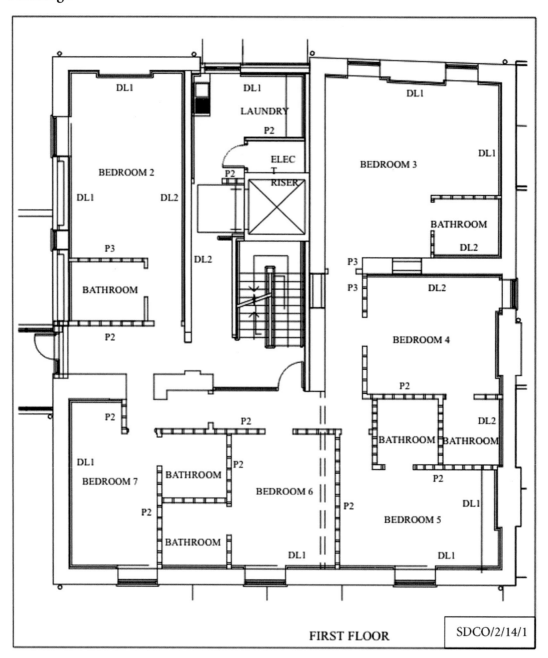

Drawing SDCO/2/14/1 Partitions first floor plan.

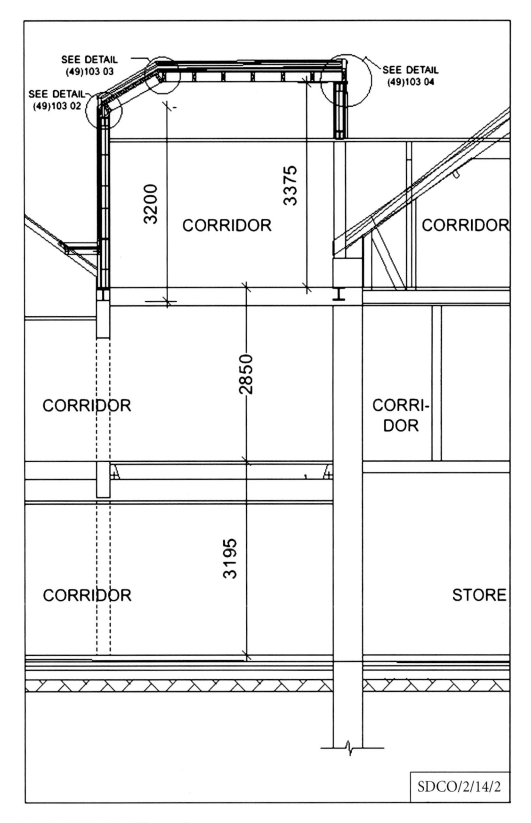

SEE DETAIL
(49)103 03

SEE DETAIL
(49)103 02

SEE DETAIL
(49)103 04

3200

3375

CORRIDOR

CORRIDOR

2850

CORRIDOR

CORRI-
DOR

3195

CORRIDOR

STORE

SDCO/2/14/2

Drawing SDCO/2/14/2 Partitions section.

Specifications

Table 14.1 Partitions specifications.

SPECIFICATION S14 PARTITIONS: HOSTEL

PARTITION SPECIFICATION No. 1

Manufacturer and reference: British Gypsum Ltd, Gypwall.
Nominal thickness (excluding finishes): 132mm
Performance criteria: Fire resistance: 60 min, Sound insulation: Rw 54db, Partition duty to BS 5234: Parts 1 and 2: Severe, Maximum height: 4900mm, Timber sole plate: Not required.
Framing: Studs: Gypframe 70 S 50 at 600mm centres, and at wall abutments, Stud boxing: Floor channel: Gypframe 72 C 50, Head channel: Gypframe 72 C 50.
Frame fixing: Gypframe GFT1 to support horizontal board joints of single layer plasterboard linings. Fixing strap: GFS1 to support horizontal joints in double layer plasterboard linings.
Lining: 2 No. 15mm layers Soundblock Board with tapered edges.
Lining fixing
Face layer: Fix securely to all supports at maximum 300mm centres (reduced to 200mm at external angles where recommended by the board manufacturer). Previous layer of plank plasterboard: Install with long edges at right angles to studs, and fix securely to each stud using two screws. Other previous layers: Fix securely to supports around the perimeter of each board at maximum 300mm centres. Fix working from the centre of each board. Position screws not less than 10mm from the edge of the board. Set heads in a depression; do not break paper or gypsum core.
Cavity insulation: 25mm Gypglass 1200.
Sealant: To be applied at all perimeters.
Finishing: Taped seamless finish.
Other requirements: Drywall Primer Coat over entire wall surface. Provide 12mm Plywood Reinforced Zone from 300 to 1500 to both sides of partition and set between studs.

PARTITION SPECIFICATION No. 2

Manufacturer and reference: British Gypsum Ltd, Gypwall.
Nominal thickness (excluding finishes): 164mm.
Performance criteria: Fire resistance: 90 min, Sound insulation: Rw 54db, Partition duty to BS 5234: Parts 1 and 2: Severe, Maximum height: 7.8mm, Timber sole plate: Not required
Framing: Studs: Gypframe 70 C 50 at 600mm centres, and at wall abutments, Stud boxing: No required, Floor channel: Gypframe 70 C 50, Head channel: Gypframe 70 C 50
Frame fixing strap: GFS1 to support horizontal joints of single layer plasterboard linings.
Lining: 2 No. 15mm Layers Soundblock Board with Tapered Edges.
Lining fixing
Face layer: Fix securely to all supports at maximum 300mm centres (reduced to 200mm at external angles where recommended by the board manufacturer). Previous layer of plank plasterboard: Install with long edges at right angles to studs, and fix securely to each stud using two screws. Other previous layers: Fix securely to supports around the perimeter of each board at maximum 300mm centres. Fix working from the centre of each board. Position screws not less than 10mm from the edge of the board. Set heads in a depression; do not break paper or gypsum core.
Cavity insulation: 50mm Gypglass 1200.
Sealant: To be applied at all perimeters.
Finishing: Taped seamless finish.
Other requirements: Drywall Primer Coat over the entire wall surface. Provide 12mm Plywood Reinforced Zone from 300 to 1500 to both K10 Plasterboard dry linings/partitions.

Table 14.1 (*Continued*)

SPECIFICATION S14 PARTITIONS: HOSTEL
PARTITION SPECIFICATION No. 3

Manufacturer and reference: British Gypsum Ltd, Gypwall.
Nominal thickness (excluding finishes): 100mm.
Performance criteria: Fire resistance: 60 min, Sound insulation: Rw 54db, Partition duty to BS 5234: Parts 1 and 2: Severe, Maximum height: 4900mm, Timber sole plate: Not required.
Framing: Studs: Gypframe 70 S 50 at 600mm centres, and at wall abutments.Stud boxing:
Floor channel: Gypframe 72 C 50, Head channel: Gypframe 72 C 50.
Fixing T: Gypframe GFT1 to support horizontal board joints of single layer.
plasterboard linings. Fixing strap: GFS1 to support horizontal joints in face layer boards of double layer plasterboard linings.
Lining: 1 No. 15mm layer Soundblock Board with tapered edges. Fixing: As clause 591A
Screws: Gyyproc Drywall.
Lining fixing
Face layer: Fix securely to all supports at maximum 300mm centres (reduced to 200mm at external angles where recommended by the board manufacturer). Previous layer of plank plasterboard: Install with long edges at right angles to studs, and fix securely to each stud using two screws. Other previous layers: Fix securely to supports around the perimeter of each board at maximum 300mm centres. Fix working from the centre of each board. Position screws not less than 10mm from the edge of the board. Set heads in a depression; do not break paper or gypsum core.
Cavity insulation: 25mm Gypglass 1200.
Sealant: To be applied at all perimeters.
Finishing: Taped seamless finish as clause 671A.
Other requirements Drywall Primer over entire wall surface. Provide 12mm Plywood Reinforced Zone from 300 to 1500 to both sides of partition and set between studs.

DRY LINING SPECIFICATION No. 1

Manufacturer and reference: British Gypsum Ltd, GypLyner wall lining.
Background: Existing Plaster/New Blockwork depending on Location, Stand off (background to face of channel): 20mm.
Framing: Grid: Gypframe GL1 channels at 600mm nominal centres, Channel connectors: Gypframe GL3, Fixing brackets: Gypframe GL2/GL9 at 800mm maximum centres, Floor/ceiling track: Gypframe GL8.
Cavity insulation: As clause 580.
Lining: 65mm Gyproc Thermal Board Super. Fixing: Securely fix GL8 floor/ceiling track at 600mm centres, Position GL2/GL9 fixing brackets at equal vertical centres and fix to background with GyprocGyplyner Anchors for solid backgrounds or proprietary fixings for hollow backgrounds.
Thermal sealant: To all penetrations.
Acoustic sealant: Required at all junctions.
Finishing: Taped seamless finish.
Other requirements: Drywall Top Coat Sealer Coat in 2 No. coats over entire wall surface.

(Continued)

Table 14.1 (*Continued*)

SPECIFICATION S14 PARTITIONS: HOSTEL

DRY LINING SPECIFICATION No. 2

Manufacturer and reference: British Gypsum Ltd GypLyner wall lining.
Background: Existing Plaster Blockwork depending on location, Stand off (background to face of channel): 20mm.
Framing: Grid: Gypframe GL1 channels at 600mm nominal centres.
Channel connectors: Gypframe Gl3, Fixing brackets: Gypframe GL2GL9 at 800mm maximum centres, Floor/ceiling track: Gypframe GL8.
Cavity insulation: None.
Lining: 12.5mm Gypsum Wall Board with Tapered Edges sheet width 1200mm. Fixing: Securely fix GL8 floor/ceiling track at 600mm centres, Position GL2/GL9 fixing brackets at equal vertical centres and fix to background with GyprocGyplyner Anchors for solid backgrounds or proprietary fixings for hollow backgrounds.
Thermal sealant: None.
Acoustic sealant: Required at all junctions.
Finishing: Taped Seamless finish.
Other requirements: Drywall Top Coat Sealer Coat in 2 No. coats over entire wall surface.

DRY LINING SPECIFICATION No. 3

Manufacturer and reference: British Gypsum Ltd GypLyner wall lining.
Background: Existing Plaster Blockwork depending on location, Stand off (background to face of channel): 20mm
Framing: Grid: Gypframe GL1 channels at 600mm nominal centres
Channel connectors: Gypframe Gl3, Fixing brackets: Gypframe GL2GL9 at 800mm maximum centres, Floor/ceiling track: Gypframe GL8.
Cavity insulation: None.
Lining: 50mm GyprocThemal Board Super with Tapered Edges sheet width 1200mm. Fixing: Securely fix GL8 floor/ceiling track at 600mm centres, Position GL2/GL9 fixing brackets at equal vertical centres and fix to background with GyprocGyplyner Anchors for solid backgrounds or proprietary fixings for hollowbackgrounds.
Thermal sealant: None.
Acoustic sealant: Required at all junctions.
Finishing: Taped Seamless finish.
Other requirements: Drywall Top Coat Sealer Coat in 2 No. coats over entire wall surface.

Query sheet

Table 14.2 Partitions query sheet.

QUERY SHEET: PARTITIONS	
QUERY **(From the QS)**	**ANSWER** **(Assumptions are to be confirmed)**
1. Floor to ceiling height	Confirmed xx.xx.xx
2 Reveals	Assumed 400mm. Confirmed xx.xx.xx
3. Wet finishes	Included elsewhere
4. Waterproof boards to bathrooms	Confirmed xx.xx.xx
5. Sound proofing to bedrooms	To corridors only
6. Sealing at soffit	Confirmed
7. Fireproofing at suspended ceilings	Confirmed
8. Partition Specification No. 1	Ground floor only
9. Drylining specification No.3	Not used on first floor
10. Height of windows	Assumed
11. Bulkhead to staircase	Assumed
12 Finishes to underside of staircases	None
13. Steel beam to staircase	Assumed
14. Steel columns to laundry windows	Assumed

14.2   TECHNOLOGY

Dry finishes consist mainly of proprietary systems that have a detailed specification and comprehensive fixing systems. It is common for every wall in each room to have a different specification *viz.* heat insulation on external walls, water resistance in kitchens and bathrooms, fire rating against circulation space, sound insulation against adjacent rooms. The manufacturer's instructions provide all the information required in the NRM concerning construction and materials.

Finishing schedules

Finishes can be measured using spreadsheets with calculations and formulae built into the cells. A simple table is included in Table 14.3. A spreadsheet example is included in Chapter 16, Finishes, using the medical centre. An extensive finishes schedule requires a great deal of formatting and is as complex as traditional taking-off processes. Checking the final quantities is an important part of measuring finishes because of the extensive use of repeated measurements. A single error when multiplied numerous times can lead to a large error. An example of a quantities check is included in Chapter 16.

Table 14.3 Finishes spreadsheet. A spreadsheet is the easiest way of preparing a large amount of repetitive quantities. Checking the quantities becomes more important when a single error can be reflected in several different cells.

PARTITIONS				DRY LININGS			
Spec no. 2		Spec no. 3		Spec no. 1		Spec no. 2	
Bed 2 2400 300 900 100	3700	2400 100 1850	4300	7800 2/3800	15400	2/8400	16800
Laundry		2000 1600 1100	4700	3200 3600 2000	8800		
Bed 3		2100 100 1800	4000	5600 3500	9100	2200 2300	4500
Bed 4 150 1900 150 1800	4000	3800 1200 400	5400			4400 9100 2/2/200	14300
	4000						
Bed 5	2100				5100		
	2100						
Bed 6 100 900 150 3325 150 2100 150 950	7825			8500 5200 1700	15400		
	2100						
3/4200	8400						
	1000						
Lengths	35225		18400		53800		35600

Table 14.4 Practical application: Hostel first floor.

			Partitions			
			HOSTEL FIRST FLOOR			
			Drawings			
			SDCO/14/2/1 Plan First floor			
			SDCO/2/14/2 Section			
			Specification S14			
						The title page includes the trade name, the name of the contract, the full drawing schedule with dated revisions, the dated specification, the name of the measurer and the date.
			SDCO			
			May 2012			

Table 14.4 Practical application: Hostel first floor (*Continued*)

Partitions

British Gysum Ltd, Gypwall, 164mm thick partition comprising frame of 70 S 50 studs at 600 centres, 72 C 50 head and floor channels, boarding of 2 No. Layers of 15mm Soundblock Board with tapered edges, taped seamless joints, 25mm Gypglass 1200 insulation, sealant applied to all perimeters, Drywall Coatover primer, FR 90 minutes, Sound insulation 54db, duty severe, fixed in accordance with manufacturer's instructions. (Specification No. 2)

Walls of each room measured in turn on a spreadsheet. Quantities are abstracted from the spreadsheet.

		Scaled	2850
			(300)
			2550

35.23		Proprietary metal framed system to form walls, 164mm thick x 2550mm high x 37,225 mm total length. [20.1.1
2.55		

Ddt
Ditto, 164mm thick wall.   [20.1.1.1.*.1

9/	.90
	2.20

Ddt

Dimension column entries:

9/	1
4/	1
6/	1
5/	1

2/2/	1.90
	2.55
2/2/	1.80
	2.55
2/2/2/	2.10
	2.55

3.70
4.80
8.00
1.00

Right column descriptions:

Deductions are made for all openings above the void threshold of 1.00m² NRM 2 ref 20 Note 1 and add back the labours for forming the openings.

Extra over forming openings, ne 2.50m², unlined   [20.4.1.2

Ditto, angles   [20.6

Ditto, Tee junctions   [20.7.1.1

Ditto, access panels, 450 x 450.   [20.8.1

Ditto, extra over for moisture resistant MR wallboard   [20.3.1   [Bth 4/5/6/7

Ditto, fire seals.   [20.5.1.1.1.1   [Junction with corridor

Fire seals are measured in NRM 2 Section 20.5 but could also be measured in Section 31.

Table 14.4 Practical application: Hostel first floor (*Continued*)

Dim (L)	Description	Dim (R)	Description (R)
	Partitions	5/ .90 / 2.20	Ddt Ditto, 100mm thick wall. [20.10.1.*.1
	British Gysum Ltd, Gypwall, 100mm thick partition comprising frame of 70 S 50 studs at 600 centres, 72 C 50 head and floor channels, boarding of 1 No. Layer of 15mm Soundblock Board with tapered edges, taped seamless joints, 25mm Gypglas s 1200 insulation, sealant applied to all perimeters, Drywall Coatover primer, FR 60 minutes, Sound insulation 54db, duty severe, fixed in accordance with manufacturer's instructions. (Specification No. 3)	5/ 1	Extra over forming openings, ne 2.50m^2, unlined [20.4.1.2
		3/ 1	Ditto, angles [20.6
		5/ 1	Ditto, Tee junctions [20.7
		5/ 1	Ditto, access panels, 450 x 450 [20.8.1
18.40 / 2.55	Proprietary metal framed system to form walls, 100mm thick x 2550mm high x 13,700 mm total length [20.1.1	2.10 / 2.55 / 1.80 / 2.55	Ditto, extra over for moisture resistant MR wallboard [20.3.1 [Bth 2/3
2/ 1.66	Ditto, fire seals. [20.5.1.1.1.1 [Junction with corridor	2/35.23 / 1.20 / 2/18.40 / 1.20 / 53.80 / 1.20 / 35.60 / 1.20	Extra over all wall board for reinforcement zone 1200mm high to drylining and both sides of partition [20.3.1

Table 14.4 Practical application: Hostel first floor (*Continued*)

		Partitions			*Window 600 x 1200 in bedroom 2 is a void < 1.00m² and is not deducted.*
		British Gysum Ltd, GypLyner wall lining, to new walls, 20mm from base, grid of Gypframe GL1 channels at 600mm centres, GL8 head and floor channels, boarding of 65mm Thermal Board Super with tapered edges, taped seamless joint, thermal sealant applied to all openings, acoustic sealant to all junctions, 2 No. coats of Drywall Coatover primer, fixed in accordance with manufacturer's instructions. (D/L Specification No. 1)	6/	1.20 1.20 1.30 1.20	Ddt Ditto, wall lining. [20.10.1*.1
			8/	1	Extra over forming openings, ne 2.50m², unlined [20.15.1.2
			31/	1	Ditto, angles [20.17
	53.80 2.55	Proprietary linings to walls, over 300mm wide on face, Gypframe fixing to new blockwork walls, [20.10.1.*. 1	2/	1	Ditto, Tee junctions [20.18.1
	2.10 2.55 1.85 2.55	Ditto, extra over for moisture resistant MR wallboard [20.3.1 [Bth 2/6	5/	1	Ditto, fair ends PROVISIONAL [20.10.1.1.1.1 *Provisional quantities are an allowance for work that cannot be measured at this stage.*

Table 14.4 Practical application: Hostel first floor (*Continued*)

Partitions

		British Gysum Ltd, GypLyner wall lining, to existing walls, 20mm from base, grid of Gypframe GL1 channels at 600mm centres, GL8 head and floor channels, boarding of 12.5mm Wall Board with tapered edges, taped seamless joint, acoustic sealant to all junctions, 2 No. coats of Drywall Coatover primer, fixed in accordance with manufacturer's instructions. (D/L Specification No. 2)	1.20 1.20	Ddt Ditto, wall lining. [20.10.1*.1
			1	Extra over forming openings, ne 2.50m^2, unlined [20.15.1.2
			2/ 1	Ditto, angles [20.17
			2/ 2.50	Ditto to steel column, girth 600–800mm, to laundry [20.12.3
	35.60 2.55	Proprietary linings to walls, over 300mm wide on face, Gypframe fixing to existing walls [20.10.1.*.1	2.10	Ditto, 3–600 girth, to link [20.12.3
8/2/	1.20 .60	Ditto, ne 300mm wide, do [20.10.2	3.00	Ditto, bulkhead, 900 girth, staircase [20.14.4
7/	1.20	[Reveals & soffits to windows & Ditto, angle beads, to reveals, do [20.18	1.80 2.55 2.10 2.55	Ditto, extra over for moisture resistant MR wallboard [20.3.1 [Bth 3/4
			End of partitions	

14.4 SELF-ASSESSMENT EXERCISE: HOSTEL GROUND FLOOR

Measure the partitions and linings to the ground floor as shown on Drawing SDCO/2/E14/1 in Appendix 14 and Drawing SDCO/2/14/2. Please prepare your own measurement using a spreadsheet and blank double dimension paper in Appendix 1a and also prepare a query sheet of problems that you have encountered. Compare your own work with the proposed solution in Appendix 14 (Table E14.2) and self-assess your work on the assessment sheet in Appendix 1b.

To provide further assistance there are dedicated websites at http://ostrowski quantities.com and at Wiley Blackwell (http://www.wiley.com/go/ostrowski/measurement). It is hoped that the provision of this will go some way towards explaining the concepts and principles more clearly than using the printed word alone.

15 Curtain Walling

15.1 Measurement information
  - Drawings
  - Specification
  - Query sheet
15.2 Technology
15.3 Practical application: Aircraft showroom gridline 1
15.4 Self-assessment exercise: Gridline A

15.1   MEASUREMENT INFORMATION

Drawings

See Drawings SDCO/2/15/1 (curtain walling plan and details) and SDCO/2/15/2 (curtain walling details).

Measurement Using the New Rules of Measurement, First Edition. Sean D.C. Ostrowski.
© 2013 John Wiley & Sons, Ltd. Published 2013 by John Wiley & Sons, Ltd.

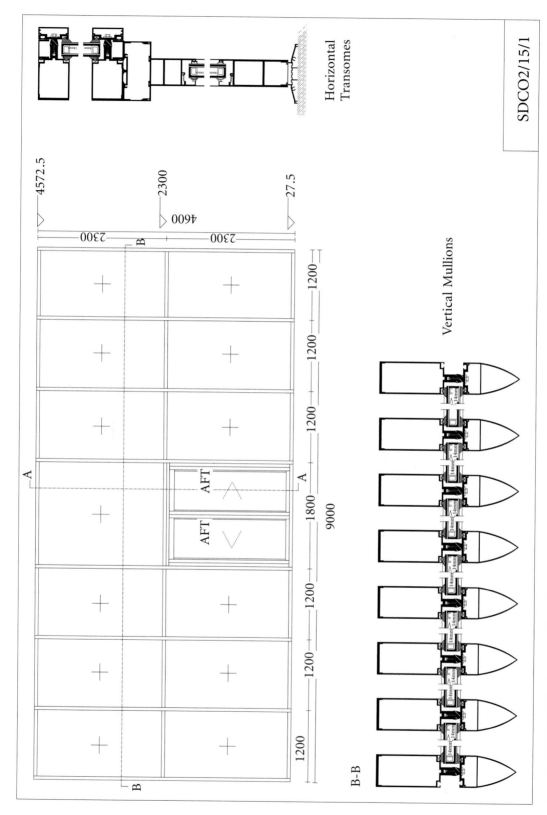

Drawing SDCO/2/15/1 Curtain walling plan and details.

SDCO2/15/1

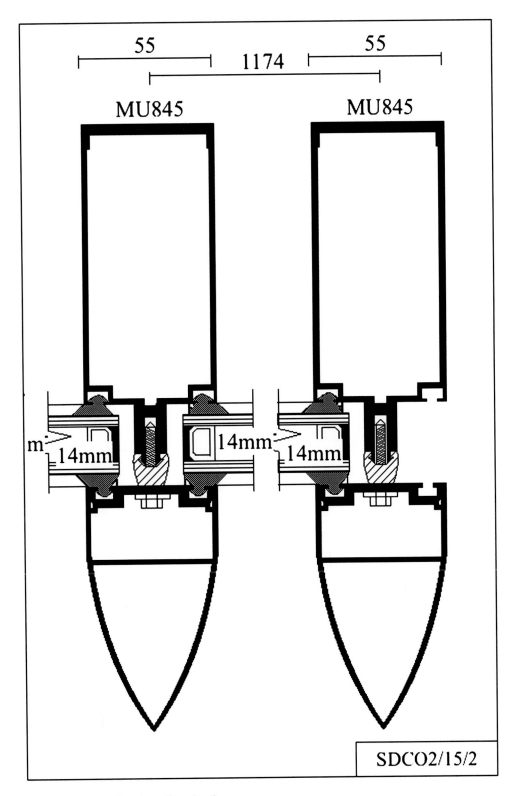

Drawing SDCO/2/15/2 Curtain walling details.

Specification

Table 15.1 Curtain walling specification.

Specification S15 Curtain walling
Techmal MX system or equal and approved
Extrusions to BS EN 12020-2
Anodised to BS 1615
Powder coating. Organic to BS 6496
Mullions MU845/847D/889
Transoms MU 846/870/850
Double glazing unit, 30mm. Laminated 9mm inner pane and low-E coated. Annealed clear 8mm inner pane
Neoprene extruded gaskets
Colour co-ordinated beads, mechanically fixed.

Query sheet

Table 15.2 Curtain walling query sheet.

QUERY SHEET	
OPENINGS AND WINDOWS	
QUERY (From the QS)	ANSWER (From the Architect/Engineer) (Assumptions are to be confirmed by QS)
1.Fixing detail to steelwork 2. Infill panel specification 2. Assumed Schuco stick system 3. Security system for door 4. Glass specification	Assumed xx.xx.xx Assumed xx.xx.xx No. xx.xx.xx Confirmed xx.xx.xx Assumed xx.xx.xx

15.2   TECHNOLOGY

Small areas of curtain walling can be installed using a 'stick' system. This is a framework of mullions and transoms that is fixed to the structure and houses the glazing. Larger areas that can provide the full cladding system for a high building use a subframe attached to the structure.

The glazing units are sealed double or triple glazed units with different specifications for each pane of glass. They are fixed to the glazing framework with mechanical fixings and profiled neoprene gaskets. The glazing framework is usually profiled aluminium which often includes decorative fittings and finishes and is mechanically fixed to the subframe with proprietary fixings. The subframe is usually a steel framework bolted to the structure. If the structure is reinforced concrete this will require anchor bolts fixed into the concrete as the framework is being constructed.

Erecting the glazing panels requires access for the fixers on the internal elevation and on the outside elevation where access is required between the scaffolding and the outside face of the building. Site storage is unlikely due to the risk of breakage and the need for substantial crane time to double handle the unit into storage and out of storage to the workface. Erection therefore takes place immediately on arrival of the panel and requires a dedicated crane to provide the vertical and horizontal movement necessary.

The glass, gaskets and framing are usually made in separate factories and assembled as a prefabricated glazing panel and then transported to site. The subframe is often a separate work package.

The provision of various kinds of doors, ventilation panels, access panels, automatic opening vents and the like requires substantial amendments to the standard design and may require a unique design and manufacturing solution.

Table 15.3 Practical application: Aircraft showroom gridline 1.

			Curtain walling AIRCRAFT SHOWROOM Drawings SDCO/2/15/1 Elevation SDCO/2/15/2 Details Specification S15 SDCO May 2012				*The title page includes the trade name, the name of the contract, the full drawing schedule with dated revisions, the dated specification, the name of the measurer and the date.*

Table 15.3 Practical application: Aircraft showroom gridline 1 (*Continued*)

		Curtain walling			
		Techmal MX Trame glazing system comprising aluminium extrusions to BS EN 1202-2, anodised to BS 1615, and powder organic coated to BS 6496, mullion components MU845/847D/889 at 1200mm centres, transom components MU846/870/850 complete with gaskets, sealant and mechanical assembly as drawing No. MU/XXX Typical Construction Details; 30mm thick sealed double glazed units with 9mm lo-E coated laminated inner pane, 14mm cavity, 8mm clear annealed outer pane, fixed with extrudeded neoprene gaskets and beads. Fixed to existing steel frame using mild steel brackets and cleats and mechanical fixing, to gridline A	2/ 1		Ditto, extra over cover panel to beams, vertical, external, insulated and finished as shown on drawing No. DETL XXXX Rev.04, 150 wide x 4600 high [21.8.1.2
			1		Ditto, extra over, pair of doors, 1800 x 2300 overall, additional mullion and transom components MS637 [21.8.1.2
			9.00		Ditto, boundary work, aluminium flashing, heads, horizontal [21.9.1.1 & Bottom edge, aluminium cill plate, horizontal [21.9.1.1
9.00 4.60		Walls, exceeding 600 wide, vertical, external [21.1.2.3.2	2/ 4.60		Ditto, abutments, flashing, vertical [21.9.8.2

End of curtain walling

15.4　SELF-ASSESSMENT EXERCISE: GRIDLINE A

Measure the curtain walling to gridline 1 using the elevation on Drawing SDCO/2/E15/1 in Appendix 15. Please prepare your own measurement using blank double dimension paper in Appendix 1a and also prepare a query sheet of problems that you have encountered. Compare your own work with the proposed solution in Appendix 15 (Table E15.1) and self-assess your work on the assessment sheet in Appendix 1b.

To provide further assistance there are dedicated websites at http://ostrowski quantities.com and at Wiley Blackwell (http://www.wiley.com/go/ostrowski/measurement). It is hoped that the provision of this will go some way towards explaining the concepts and principles more clearly than using the printed word alone.

16 Finishes

16.1 MEASUREMENT INFORMATION

Drawings

See Drawings SDCO/2/16/1 (finishes plan) and SDCO/2/16/2 (finishes elevation).

Measurement Using the New Rules of Measurement, First Edition. Sean D.C. Ostrowski.
© 2013 John Wiley & Sons, Ltd. Published 2013 by John Wiley & Sons, Ltd.

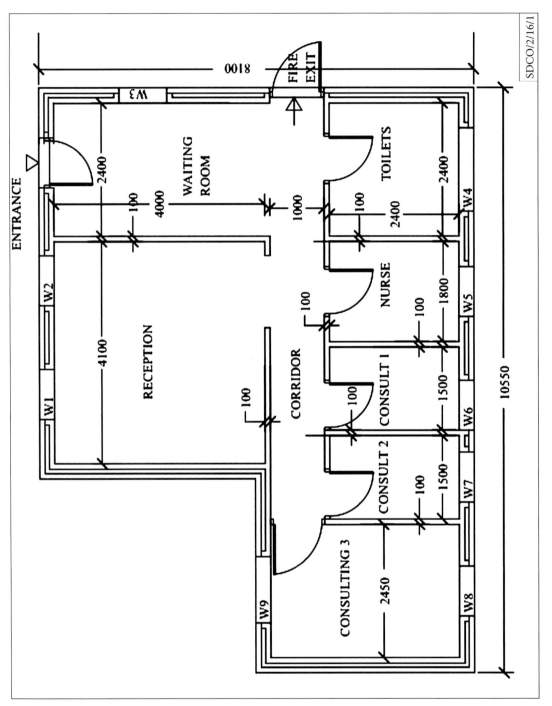

Drawing SDCO/2/16/1 Finishes plan.

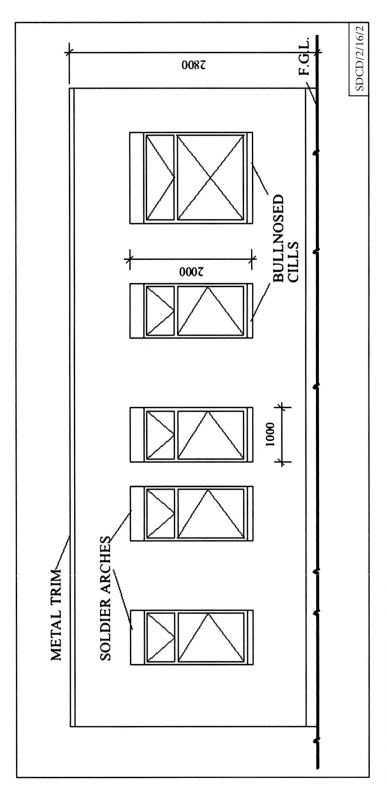

Drawing SDCO/2/16/2 Finishes elevation.

Specification

Table 16.1 Finishes specification.

SPECIFICATION S15 FINISHES: MEDICAL CLINIC				
ROOM	**FLOOR**	**SKIRTING**	**WALL**	**CEILING**
GENERAL	75mm Sand and cement screed		11mm Two coats render & set Mist and two coats emulsion	7mm Plasterboard & 2mm, set coat. Mist and two coats emulsion
WAITING ROOM	5mm Vinyl Sheet	100mm coved	Ditto	
CORRIDOR	5mm Vinyl Sheet	100mm coved	Ditto	
RECEPTION	Carpet tiles (PC £25/m²)	100 × 25 Softwood Knot, prime and two coats gloss	Ditto	
TOILETS	150 × 150 × 22mm thick Red quarry tiles	150mm Bullnose Red quarry tiles	Ditto plus 100 × 100 × 5mm White glazed wall tiles to 1.20m above floor level	
NURSE	Carpet tiles (PC £25/m²)	100 × 25 Softwood Primed and two coats gloss	Ditto plus 100 × 100 × 5mm White glazed wall tiles to 1.20m above floor level	
CONSULTING 1/2/3	Carpet and underlay (PC £50/m²)	100 × 25 Softwood Primed and two coats gloss	Ditto plus 75 × 25mm softwood picture rail	
	Carpet trim, 25 × 5mm Brass dividing strip			

Query sheet

Table 16.2 Finishes query sheet.

QUERY SHEET: FINISHES	
QUERY (From the QS)	**ANSWER** (Assumptions are to be confirmed)
1. Internal height of wall. 2. Floor screed 3. Specification for angle bead 4. Standard colours	Confirmed xx.xx.xx Assumed 75mm. Confirmed xx.xx.xx Confirmed xx.xx.xx Confirmed xx.xx.xx

16.2 TECHNOLOGY

Wet finishes have a wide variety of specifications depending on the background to which they are applied and the function they will serve.

Dry finishes consist mainly of proprietary systems that have a detailed specification and comprehensive fixing systems.

Finishing schedules

Finishes are best completed using spreadsheets with calculations and formulae built into the cells. Examples using the medical centre in the Practical application are given in Tables 16.4, 16.5, 16.6 and 16.7. An extensive finishes schedule requires a great deal of formatting and is as complex as traditional taking-off processes.

Checking the final quantities is an important part of measuring finishes because of the extensive use of repeated measurements. A single error when multiplied numerous times can lead to a large error. An example of a quantities check is included in Table 16.8.

16.3 PRACTICAL APPLICATION: MEDICAL CENTRE

Table 16.3 Practical application: Medical centre.

			Finishes				
			MEDICAL CENTRE				
			Drawings				
			SDCO/16/2/1 Plan				
			SDCO/16/2/2 Elevation				
			Specification S15				*The title page includes the trade name, the name of the contract, the full drawing schedule with dated revisions, the dated specification, the name of the measurer and the date.*
			SDCO May 2012				

Table 16.3 Practical application: Medical centre (*Continued*)

Finishes

Dimension check

N–S		E–W	
2/250 ✓	500 ✓	2/250 ✓	500 ✓
	4000 ✓		2450 ✓
2/100 ✓	200 ✓	4/100 ✓	400 ✓
	1000 ✓	2/1500 ✓	3000 ✓
	2400 ✓		1800 ✓
	8100 ✓		2400 ✓
			10550 ✓

Wall finishes

External   2800
Floor screed 75
Roof parapet
Say   275
100
50   (500)
Internal wall height 2300

Waiting room

100
1000
4000   5100 x 2400

Corridor

100
1800
100
1500
100
1500
100   5200 x 1000

Consulting 3

2400
100
1000
2450 x 3500

Internal wall height is calculated from the external dimensions.

Length and breadth of each room calculated in advance.

A 'to-take' list can act as a check for measuring different specifications.

	To take list
Floor	finishes
	Screed
	Tiling
	Carpet tiles
	Carpet & underlay
	PVC
	Skirtings
	Threshold
	Dividing strip
Wall	finishes
	Plaster
	Tiling
	Picture rails
	Deduct openings
Ceiling	finishes
	Plaster
Decorations	

Table 16.3 Practical application: Medical centre (*Continued*)

Finishes

		Finish to walls 13mm two coat render and set, > 600 wide, to blockwork	*Walls of each room measured in turn.*
2/	2.40		
	2.30		[28.7.2
2/	5.10	11.04	*Because finishes generates extensive dimensions where a new column appears for the same item the description should be repeated.*
	2.30	23.46	[Waiting room
2/	1.00		
	2.30	4.60	
2/	5.20		[Corridor
	2.30	23.92	
2/	4.10		
	2.30	18.86	
2/	4.00		
	2.30	18.40	[Reception
2/2/	2.40		
	2.30	22.08	
2/	1.80		
	2.30	8.28	[Toilets
2/	2.40		
	2.30	11.04	[Nurse
2/	1.50		
	2.30	6.90	
2/	2.40		
	2.30	11.04	
		159.62	

Right-hand dimension column:

			Walls 13mm two coat plaster render and set, 600 wide, to blockwork
2/	1.50		[28.7.2
	2.30	6.90	
2/	2.40		[Consulting 2
	2.30	11.04	
2/	3.50		[Consulting 3
	2.30	16.10	
2/	2.45		
	2.30	11.27	
		45.31	

Ddt ditto

Ddt			
2/1.00			
2.20	(4.40)		[Recept'n
7/2/ .80			
2.20	(24.64)		Doors
7/1.00			
2.00	(14.00)		[W1,2,3,5,6,7,8
2/1.50			
2.00	(6.00)		[W4,9
	(49.04)		

Table 16.3 Practical application: Medical centre (*Continued*)

Finishes

Ceramic tiles, walls, plain, > 600 wide.
[28.7.2
[Nurses

2/ 4.20
1.20 10.08
2/ 4.80
1.20 11.52
21.60

[Toilet

R & S 159.62
45.31
(49.04)
155.89

By 'squaring' the totals for plastering and tiling it provides simple dimensions for the decorations.

155.89
Ddt 1.00
2/1.60   155.89
1.00
(21.60)
134.29

Painting to general surfaces, walls, > 300 girth, internal, one mist and two coats emulsion
[29.1.2.1

Ceilings

Ceilings, 7mm plasterboard > 600 wide, to concrete
[28.9.2

&

Ceilings, 2 mm one coat plaster set, > 600 wide, to plasterboard
[28.6.3
[Waiting

4.10
2.40 9.84
7.60
1.00 7.60

[Corridor

4.10
4.00 16.40

Reception

2.40
2.40 5.76
1.80

[Toilets

2.40 4.32

[Nurse

2/ 1.50
2.40 7.20
3.50

[Consulting 1/2

2.45 8.58

[Consulting 3

59.70

Painting to general surfaces, ceilings, >300 gth, internal, one mist and two coats emulsion
[29.1.2.1

59.70
1.00 59.70

Table 16.3 Practical application: Medical centre (*Continued*)

Finishes

Screed

Floor screed covers the whole floor so are measured across partitions.

Gross area

E–W		N–S
10550		8100
2/250 (500)		(500)
10050		7600

Deduct area outside building

2450	4000	
100	100	4100
Say 1000		
3550		

Ddt	10.05	
3.55	7.60	
4.10	76.38	Screed, sand and cement (1:10), to concrete, 75mm th, >600 wide, level and to fall, <15°
	(14.56)	[28.1.2.1
	61.83	

Floors

2.40		Floors, > 600 wide, quarry tiles, plain, 22 x 150 x 150, level or to falls, <15°
2.40	5.76	[28.2.2.1 [Toilets
2/2/ 2.40	9.60	Ditto, skirting, 150mm high, bullnose
Ddt		
.80	(.80)	[28.14.1.*.1.
	8.80	
5.20		Floors, > 600 wide, red uPVC sheeting, fixed with adhesive, butt joints
1.00	5.20	
2.40		
5.10	12.24	[28.2.2 [Corridor
	17.44	[Waiting room
2/ 5.20		Ditto, coved skirting, 150mm high
2/ 1.00		
2/ 2.40		[28.14.1.*.1
2/ 5.10	27.40	
Ddt		
5/ .80		
3/1.00	(11.80)	
2/2.40	15.60	

Table 16.3 Practical application: Medical centre (*Continued*)

Finishes

Floors

Floors, > 600mm wide, carpet tiles (PC £25/m²)
[28.2.2
[Reception

[Nurse

```
   4.10
   4.00
   1.80    16.40
   2.40
           4.32
          20.72
```

Ditto, > 600mm wide, carpet including underlay (PC £50/m²)
[28.2.2
[Consulting 1/2

[Consulting 3

```
2/  1.50
    2.40    7.20
    3.50
    2.45    8.57
           15.77
```

Picture rail, 75 x 25, moulded softwood, to blwk
[22.2.1

```
2/2/  1.50
2/2/  2.40
2/    3.50
2/    2.45
             27.50
```

End of finishes

Floors

Skirting, 150 x 25, softwood, to blockwork
[22.1.1.

[Reception

[Consulting 1/2

[Nurse

[Consulting 3

[Door openings

```
2/   4.10   8.20
2/   4.00   8.00
2/2/ 1.50   6.00
2/2/ 2.40   9.60
2/   1.80   3.60
2/   2.40   4.80
2/   3.50   7.00
2/   2.45   4.90
           52.10
```

Ddt
```
2/5/   .80
      1.00
        (9.00)
        43.10
```

Painting to GS, girth < 300, internal, KPS & ②, colour
[29.1.1.

```
   27.50
   43.10
          70.60
```

Dividing strip, 25 x 15mm, p & s to screed, brass
[28.27.1

```
7/   .80
          5.60
```

Table 16.4 Finishes dimensions.

SDCO/Finishes						May 2012
MEDICAL CENTRE FINISHES SCHEDULE 1. DIMENSIONS						
	VERTICAL		**N–S**		**E–W**	
Internal Room Height			2 × 250	500	2 × 250	500
External height	Scaled	2800		4000		2450
Floor finishes	Assumed	75	2 × 100	200	4 × 100	400
Roof	Assumed			1000	2 × 1500	3000
Parapet		275		2400		1800
Roof slab		100		8100		2400
Roof coverings		50 (500) 2300				10550

					MEDICAL CENTRE FINISHES SCHEDULE 2. FLOOR FINISHES			
ROOM	**FLOOR FINISHES**							
DESCRIPTION	DIMENSIONS			Screed 75mm thick	225 × 225 × 3mm red Upvc sheeting	10mm Carpet (PC £25 per m2)	Carpet & U/L (PC £50 per m2)	22 × 150 × 150mm red quarry tiles
WAITING ROOM	Figured		2400					
	Figured	100						
	Figured	1000						
	Calculated	4000	5100	12.24	12.24			
CORRIDOR	Figured		1000					
	Figured	100						
	Figured	1800						
	Figured	100						
	Figured	1500						
	Figured	100						
	Figured	1500						
	Figured	100	5200	5.20	5.20			
RECEPTION	Figured	4100						
	Calculated	4000		16.40		16.40		
TOILETS	Figured	2400						
	Figured	2400		5.76				5.76
NURSE	Figured	1800						
	Figured	2400		4.32		4.32		
CONSULTING 1	Figured	1500						
	Figured	2400		3.60			3.60	
CONSULTING 2	Figured	1500						
	Figured	2400		3.60			3.60	
CONSULTING 3	Figured	2400						
	Figured	100						
	Figured	1000	3500					
	Calculated		2450	8.58			8.58	
SCREED	$(10.05 \times 7.60) - (3.55 \times 4.00)$			62.18	17.44	20.72	15.78	5.76
Deducts	Walls $24.30 \times .10 = 2.43$			(2.43)				
				59.75				

Table 16.5 Floor finishes.

SKIRTING							DECORATIONS
DIMENSIONS				UPVC coved skirting	25 × 100mm softwood skirting	150mm red quarry tile bull nosed skirting	Three coats gloss oil colour
2	×	2400	4800				
2	×	5100	10200	15.00	15.00		
2	×	1000	2000				
2	×	5200	10400	12.40	12.40		
2	×	4100	8200				
2	×	4000	8000	16.20		16.20	16.20
2	×	2400	4800				
2	×	2400	4800	9.60		9.60	
2	×	1800	3600				
2	×	2400	4800	8.40		8.40	8.40
2	×	1500	3000				
2	×	2400	4800	7.80		7.80	7.80
2	×	1500	3000				
2	×	2400	4800	7.80		7.80	7.80
2	×	3500	7000				
2	×	2450	4900	11.90		11.90	11.90
				27.40	52.10	9.60	52.10
			Openings	(11.80)	(9.00)	(0.80)	(9.00)
				15.60	43.10	8.80	43.10

Table 16.6 Wall finishes.

MEDICAL CENTRE FINISHES SCHEDULE 3. WALL FINISHES

NOTE – floor finish 75mm overall

DESCRIPTION	ROOM DIMENSIONS					WALL FINISHES DIMENSIONS					13mm Plaster	DIMENSIONS	100 × 100 x5mm coloured glazed wall tiles to 1.2m above floor level	DECORATIONS Two coats emulsion
WAITING ROOM	Figured		2400	×	2									
	Figured	100												
	Figured	1000												
	Calculated	4000	5100	×	2	10200	×	15.00	×	2.30	34.50	34.50		34.50
CORRIDOR	Figured		1000	×	2	2000								
	Figured	100												
	Figured	1800												
	Figured	100												
	Figured	1500												
	Figured	100												
	Figured	1500												
	Figured	100	5200	×	2	10400	×	12.40	×	2.30	28.52	28.52		28.52
RECEPTION	Figured	4100		×	2	8200								
	Calculated	4000		×	2	8000	×	16.20	×	2.30	37.26	37.26		37.26

Table 16.6 (*Continued*)

Item	Basis																
TOILETS	Figured	2400		×	2	4800											
	Figured	2400		×	2	4800	9.60	×	2.30	22.08	22.08	9.60	×	1.20	11.52	22.08	10.56
																(11.52)	
NURSE	Figured	1800		×	2	3600											
	Figured	2400		×	2	4800	8.40	×	2.30	19.32	19.32	8.40	×	1.20	10.08	19.32	9.24
																(10.08)	
CONSULTING 1	Figured	1500		×	2	3000											
	Figured	2400		×	2	4800	7.80	×	2.30	17.94	17.94						17.94
CONSULTING 2	Figured	1500		×	2	3000											
	Figured	2400		×	2	4800	7.80	×	2.30	17.94	17.94						17.94
CONSULTING 3	Figured	2400		×	2	4800											
	Figured	100	3500	×	2	7000											
	Figured	1000	2450	×	2	4900											
	Calculated	2450		×	2		11.90	×	2.30	27.37	27.37						27.37
	Deduct openings									(49.04)	(49.04)						(49.04)
											155.89				21.60		134.29
											156m²				**22 m²**		**134m²**

Table 16.7 Ceiling finishes.

MEDICAL CENTRE FINSHES SCHEDULE 4 CEILING FINISHES						
ROOM	CEILING FINISHES					DECORATIONS
DESCRIPTION	DIMENSIONS				7mm Plasterboard & 2mm skim	Three coats emulsion
WAITING ROOM	Figured		2400			
	Figured	100				
	Calculated	4000	4100	9.43	9.84	9.84
CORRIDOR	Figured		1000			
	Figured	2400				
	Figured	100				
	Figured	1800				
	Figured	100				
	Figured	1500				
	Figured	100				
	Figured	1500				
	Figured	100	7600	7.60	7.60	7.60
RECEPTION	Figured	4100				
	Calculated	4000		16.40	16.40	16.40
TOILETS	Figured	2400				
	Figured	2400		5.76	5.76	5.76
NURSE	Figured	1800				
	Figured	2400		4.32	4.32	4.32
CONSULTING 1	Figured	1500				
	Figured	2400		3.60	3.60	3.60
CONSULTING 2	Figured	1500				
	Figured	2400		3.60	3.60	3.60
CONSULTING 3	Figured	2400				
	Figured	100				
	Figured	1000	3500			
	Calculated		2450	8.58	8.58	8.58
					59.70	59.70
					60m²	**60m²**

Table 16.8 Finishes quantities check.

SDCO/Finishes										
MEDICAL CENTRE FINISHES SCHEDULE 5 QUANTITIES CHECK										
FLOOR FINISHES					**WALL FINISHES**				**CEILING**	
Screed		61.83	Skirting	52.10	R & S		155.90		Pb'd & S	59.70
			KPS & 2	52.10	Decorations	134.29			1m & 2E	59.70
UPVC	17.44				Tiling	21.60	155.89		+ Ptns 2.80	62.40
Carpet tile	20.72									
Carpet	15.78									
Quarries	5.76									
Partitions 28m × 100mm	2.80	62.50								
The quantities for the various specifications correspond.										

16.4 SELF-ASSESSMENT EXERCISE: FLOOR, WALL AND CEILING FINISHES

Calculate the floor, wall and ceiling finishes to the waiting room, toilet and consulting room 2 as shown on Drawings SDCO/2/16/1 and SDCO/2/16/2. Please prepare your own measurement using blank double dimension paper in Appendix 1a or on a spreadsheet and also prepare a query sheet of problems that you have encountered. Compare your own work with the proposed solution in Appendix 16 (Table E16.1) and self-assess your work on the assessment sheet in Appendix 1b.

To provide further assistance there are dedicated websites at http://ostrowski quantities.com and at Wiley Blackwell (http://www.wiley.com/go/ostrowski/ measurement). It is hoped that the provision of this will go some way towards explaining the concepts and principles more clearly than using the printed word alone.

17 Drainage

17.1 Measurement information
- Drawings
- Specification
- Query sheet

17.2 Technology
17.3 Practical application: West wing
17.4 Self-assessment exercise: Drainage to patio and drive

17.1 MEASUREMENT INFORMATION

Drawings

See Drawing SDCO/2/17/1 (drainage plan: west wing).

Measurement Using the New Rules of Measurement, First Edition. Sean D.C. Ostrowski.
© 2013 John Wiley & Sons, Ltd. Published 2013 by John Wiley & Sons, Ltd.

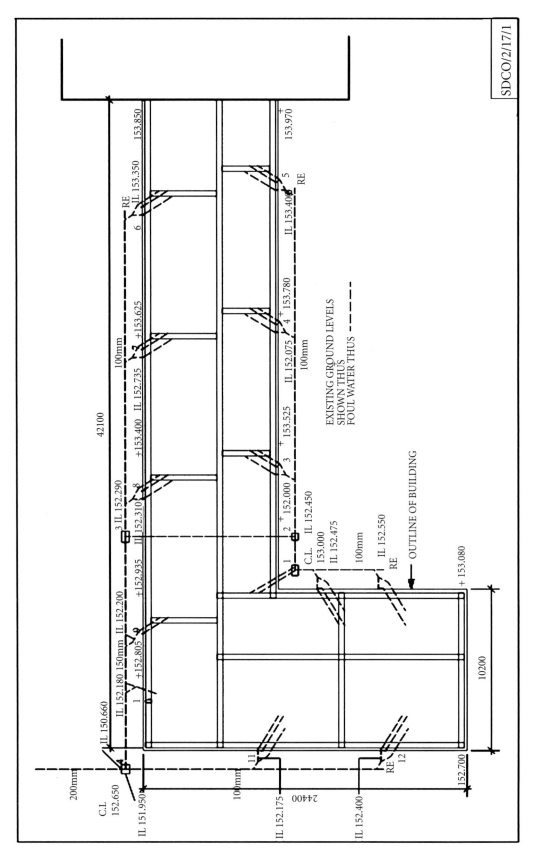

Drawing SDCO/2/17/1 Drainage plan: west wing.

Specification

Table 17.1 Drainage specification.

Specification S17 Drainage
Pipework outside the building to be uPVC, BS 4600, ring seal couplings.
Pipework inside the building to be cast iron, BS 437, coated, mechanical coupling.
Pipework outside the perimeter of the building to be bedded and surrounded with pea shingle, 150mm thick.
Pipework inside the perimeter of the building to be bedded and surrounded with concrete, Grade C10, 150mm thick.
PCC, 100 × 100mm, lintels for pipework passing under wall.
Inspection chambers to be UPVC, 600 × 250mm diameter.
Manholes to be PCC with cast iron covers and frames, complete with step irons, keys, brickwork base and raising pieces, bedded and haunched in c/m 1:3. For manholes ne 900mm total depth, internal size 450 × 800. Ditto ne 1200mm depth, size 600 × 840. Ditto ne 1500mm depth, size 690 × 1030.

Query sheet

Table 17.2 Drainage query sheet.

QUERY SHEET	
DRAINAGE	
QUERY **(From the QS)**	**ANSWER** **(From the Architect/Engineer)** **(Assumptions are to be confirmed by QS)**
1. West wing only	Confirmed xx.xx.11
2. Connections to existing services. Assumed	Confirmed xx.xx.xx
3. Manhole numbers	Confirmed xx.xx.xx
4. Short lengths?	Confirmed xx.xx.xx
5. Cast iron below buildings?	Confirmed xx.xx.xx
6. RE2-m/h 2	Confirmed xx.xx.xx
7. Gullies/Inspection chambers	Not measured
8. Raising pieces	Provisional quantities
9. iv. 153060 not 153000	Confirmed xx.xx.xx
10. Bed and surround (B & S) means 150 bed + 25mm surround	Confirmed
11. Assumed falls for branches	Confirmed

17.2   TECHNOLOGY

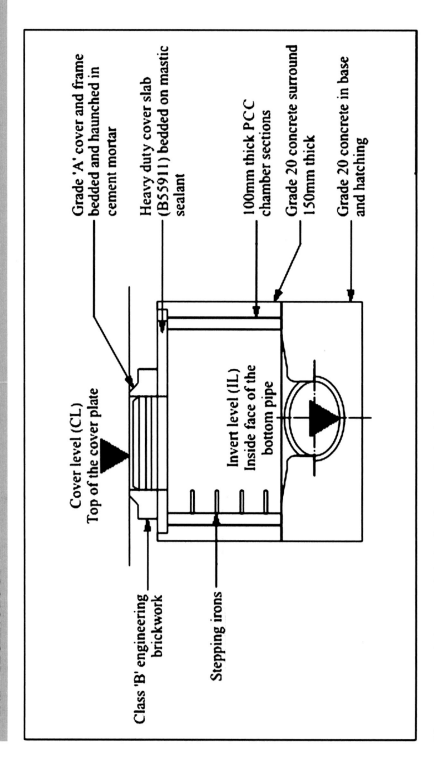

Diagram 17.1 Typical PCC manhole details. The invert level is the lowest point of the inside face of the pipework.

Table 17.3 Practical application: Drainage.

External works		
WEST WING		
DRAINAGE		
Drawings		
SDCO/2/17/1 Plan		
SDCO/2/17/2 Site plan		
SDCO/2/17/3 West wing plan		
SDCO/2/17/4 West wing plan, details		
Specification S17		*The title page includes the trade name, the name of the contract, the full drawing schedule with dated revisions, the dated specification, the name of the measurer and the date.*
SDCO May 2012		

Table 17.3 Practical application: Drainage (*Continued*)

Drainage

Drainage is most easily measured using a spreadsheet to prepare a pipework and a manhole schedule for all the work involved.

Refer to the following manhole schedule for column numbers.

Manholes are measured using NRM 2 by using the depth of the manhole from cover level, the top of the manhole, to invert level and the internal size of the manhole.

- Ascertain or calculate the existing ground level (EGL) if not provided (Column 2).
- Calculate the cover level (CL) of the top of the manhole cover if not provided (Column 3).
- Identify the IL (Column 4).
- Calculate the depth of each manhole by subtracting the IL from the CL (Column 5).
- Identify the internal size of the manhole (Column 7–12).

The NRM does not require the measurement of the total excavation from the EGL to the cover level or the external size of the manhole. However the estimator will need to calculate this in order to be able to calculate the total excavation, remove surplus spoil, planking and strutting, etc. The information to do this is included in the manhole schedule and should be provided in the Bills of Quantities and is shown in the manhole schedule (Table 17.4) below.

Table 17.4 Manhole schedule.

DRAINAGE MANHOLE SCHEDULE

M/H	EGL	CL	IL	DEPTH	PLUS BED 225mm	SIZES						EXC'N	P & S	CHANNEL		BRANCHES		ENDS	
						Internal 450 x 600	External 600 x 7500	Internal 600 x 840	External 750 x 990	Internal 690 x 1030	External 840 x 1180	m³	m²	STRAIGHT	CURVED	100	150	100	150
1	153.525	153.000	152.475	1.050	1.275						✓	1.26	5.17		1	2		4	
2	153.525	153.000	152.450	1.075	1.300						✓	1.29	5.25	1		3		3	
3	153.400	153.000	152.310	1.090	1.315						✓	1.31	5.33	1		3		2	1
4	152.700	152.650	151.950	0.750	0.975				✓			0.73	3.41		1	3		2	1
												4.588	19.166	2	2	11		11	2
4	Backdrops for 200mm pipeworkoutstanding.																		
NRM																			
1		153.000	152.475	0.525						✓					1	2		4	
2		153.000	152.450	0.550						✓				1		3		3	
3		153.000	152.310	0.690				✓						1		3		2	1
4		152.650	151.950	0.700											1	3		2	1
														2	2	11		11	2

Table 17.3 Practical application: Drainage (*Continued*)

Drainage

Drainage pipework is most easily measured using a spreadsheet to prepare a drainage schedule for all the work involved.

Refer to the pipework schedule in Tables 17.5–17.9 for column numbers.

Drainage can be measured using the NRM by using three stages for this example.

STAGE 1 DEPTHS
- Find the highest IL and start at the highest end of the pipe run
- Identify and label each manhole (Column 1)
- Identify and label both ends of all pipe runs and branches (Column 1)
- Measure the length of each pipe run (Column 2)
- Calculate the depth of each pipe run (see below)

CACULATING THE DEPTH OF EACH PIPE RUN
- Ascertain or calculate the existing ground level (EGL) (Column 5)
- Identify the invert level (IL) at the highest point, the head (Column 6)
- Calculate the depth of excavation at the head by subtracting the IL from the EGL (Column 7)
- Identify the invert level (IL) at the lowest point, the foot (Column 8)
- Calculate the depth of excavation at the foot by subtracting the IL from the EGL (Column 9)
- Calculate the average between the head and foot (Column 10)
- Add the depth of the bed and surround (Column 11)

STAGE 2 PIPEWORK OUTSIDE THE BUILDING (PVC PIPEWORK)
- Put each pipe run into the appropriate NRM depth category (Column 12–15)
- Put each pipe run into the appropriate NRM diameter category (Columns 16–21)
- Add labours for gullies, bends, junctions, reducers, rodding eye, etc. (Columns 22–28)

STAGE 3 PIPEWORK BELOW THE BUILDING (CAST IRON PIPEWORK)
- Put each pipe run into the appropriate NRM depth & diameter category (Columns 29–31)
- Add labours for gullies, bends, junctions, reducers, rodding eye, etc. (Columns 32–34)

Table 17.5 Pipework schedule 1.

DRAINAGE PIPE RUN: EXCAVATION SCHEDULE: NRM Part 1

1	2	3	4	5	6	7	8	9	10	11	12	13	14	15	16	17	18	19	20	21	22	23	24	25	26	27	28	29	30	31	32	33
					STAGE 1 DEPTHS						DEPTHS				DEPTHS: 100mm dia.			DEPTHS: 150mm dia.			STAGE 2 PVC							STAGE 3 CI				
					HEAD		FOOT			PLUS BED 150mm PLUS SURROUND 25mm=												BEND		JUNCTION				DEPTHS			CI	
M/H	LENGTH			EGL	IL	DEPTH	IL	DEPTH	AVE		.50m	<1.00m	<1.50m	<2.00m	.50m	<1.00m	<1.50m	<1.00m	<1.50m	<2.00m	GULLEY	100	150	100	150	REDUCER	RE	500	<1.00	<1.50	BEND	CONN TO GULLEY
RE – MH1	6800	6575		153.060	152.550	0.510	152.475	0.585	0.547	0.722		6575				6575											1					
Gulley 1 – pipe	800			153.060	152.550	0.510	152.500	0.560	0.535	0.710		800				800					1			1								
Branch 1	3000			153.060	152.600	0.460	152.550	0.510	0.485	0.660		3000																	3000		1	1
Branch 2	3200			153.060	152.600	0.460	152.550	0.510	0.485	0.660		3200																	3200			1
Gulley 2 – pipe	800			153.060	152.475	0.635	152.425	0.635	0.635	0.810		800				800					1			1								
Branch 1	3000			153.060	152.600	0.460	152.550	0.510	0.485	0.660		3000																	3000		1	1
Branch 2	3400			153.060	152.600	0.460	152.550	0.510	0.485	0.660		3400																	3400			1
Branch 3	3200			153.060	152.600	0.460	152.550	0.510	0.485	0.660		3200																	3200		1	1

Table 17.6 Pipework schedule 2.

DRAINAGE PIPE RUN : EXCAVATION SCHEDULE : NRM Part 2

1	2	3	4	5	6	7	8	9	10	11	12	13	14	15	16	17	18	19	20	21	22	23	24	25	26	27	28	29	30	31	32	33	34
		STAGE 1 DEPTHS			HEAD		FOOT			PLUS BED 150mm PLUS SURROUND 25mm=	DEPTHS				DEPTHS: 100mm dia.			DEPTHS: 150mm dia.			STAGE 2 PVC GULLEY	BEND		JUNCTION (PVC)				STAGE 3 CI DEPTHS				CI	
M/H	LENGTH			EGL	IL	DEPTH	IL	DEPTH	AVE		<.50m	<1.00m	<1.50m	<2.00m	<.50m	<1.00m	<1.50m	<1.00m	<1.50m	<2.00m	GULLEY	100	150	100	150	REDUCER	RE	<500	<1.00	<1.50	BEND	CONN TO GULLEY	CONN TO WC
MH1 – MH2	2200	1750		153.060	152.475	0.585	152.450	0.610	0.598	0.773		1750				1750																	
Branch 1	4000			153.060	152.525	0.535	152.475	0.585	0.560	0.735		4000																	4000			1	1
Branch 2	3500			153.060	152.525	0.535	152.475	0.585	0.560	0.735		3500																	3500			1	1
MH2 – RE	23000	22775		153.970	153.400	0.570	152.450	1.520	1.045	1.220			22775				22775										1						
Gulley 3 –pipe	800			153.525	152.550	0.975	152.500	1.025	1.000	1.175			800				800				1			1								1	1
Branch 1	3000			153.525	152.600	0.925	152.550	0.975	0.950	1.125			3000																0	3000	1	1	1
Branch 2	3200			153.525	152.600	0.925	152.550	0.975	0.950	1.125			3200																0	3200	1	1	1
Branch 3	3000			153.525	152.600	0.925	152.550	0.975	0.950	1.125			3000																0	3000	1	1	1
Gulley 4 –pipe	800			153.780	152.975	0.805	152.925	0.855	0.830	1.005			800				800				1			1								1	1
Branch 1	3000			153.780	153.025	0.755	152.975	0.805	0.780	0.955		3000																0	3000		1	1	1
Branch 2	3200			153.780	153.025	0.755	152.975	0.805	0.780	0.955		3200																0	3200		1	1	1
Branch 3	3000			153.780	153.025	0.755	152.975	0.805	0.780	0.955		3000																0	3000		1	1	1
Gulley 5 –pipe	800			153.970	153.400	0.570	153.350	0.620	0.595	0.770		800				800					1			1								1	1
Branch 1	3000			153.970	153.450	0.520	153.400	0.570	0.545	0.720		3000																0	3000		1	1	1
Branch 2	3200			153.970	153.450	0.520	153.400	0.570	0.545	0.720		3200																0	3200		1	1	1
Branch 3	3000			153.970	153.450	0.520	153.400	0.570	0.545	0.720		3000																0	3000		1	1	1

Table 17.7 Pipework schedule 3.

DRAINAGE PIPE RUN : EXCAVATION SCHEDULE : NRM Part 3

M/H	LENGTH	(3)	(4)	STAGE 1 EGL	HEAD IL	HEAD DEPTH	FOOT IL	FOOT DEPTH	AVE	PLUS BED 150mm PLUS SURROUND 25mm=	DEPTHS <.50m	DEPTHS <1.00m	DEPTHS <1.50m	DEPTHS <2.00m	100mm dia .50m	100mm dia <1.00m	100mm dia <1.50m	150mm dia <1.00m	150mm dia <1.50m	150mm dia <2.00m	STAGE 2 PVC GULLEY	BEND 100	BEND 150	JUNCTION 100	JUNCTION 150	REDUCER	RE	DEPTHS <500	DEPTHS <1.00	DEPTHS <1.50	STAGE 3 CI BEND	CONN TO GULLEY	CONN TO WC
MH2 – MH3	13000	12550		153.060	152.550	0.510	152.310	0.750	0.630	0.805		12550						12550															
MH2 – RE	21200	20975		153.850	153.350	0.500	152.290	1.560	1.030	1.205			20975						20975								1						
Gulley 6 – pipe	800			153.850	153.350	0.500	153.300	0.550	0.525	0.700		800				800	0				1			1									
Branch 1	3000			153.850	153.400	0.450	153.350	0.550	0.475	0.650		3000				0												0	3000		1	1	1
Branch 2	3200			153.850	153.400	0.450	153.350	0.550	0.475	0.650		3200				0												0	3200			1	1
Branch 3	3000			153.850	153.400	0.450	153.350	0.550	0.475	0.650		3000				0												0	3000		1	1	1
Gulley 7 – pipe	800			153.625	152.735	0.890	152.685	0.940	0.915	1.090			800			0	800				1			1									
Branch 1	3000			153.625	152.785	0.840	152.735	0.890	0.865	1.040			3000				0											0	3000	1	1	1
Branch 2	3200			153.625	152.785	0.840	152.735	0.890	0.865	1.040			3200				0											0	3200		1	1
Branch 3	3000			153.625	152.785	0.840	152.735	0.890	0.865	1.040			3000				0											0	3000	1	1	1
Gulley 8 – pipe	800			153.400	153.310	0.090	153.260	0.140	0.115	0.290	800				800	0	0				1			1									
Branch 1	3000			153.400	153.360	0.040	153.310	0.090	0.065	0.240	3000				0												3000			1	1	1
Branch 2	3200			153.400	153.360	0.040	153.310	0.090	0.065	0.240	3200				0												3200				1	1
Branch 3	3000			153.400	153.360	0.040	153.310	0.090	0.065	0.240	3000				0												3000			1	1	1
MH3 – MH4	15200		14750	152.935	152.290	0.645	150.550	2.385	1.515	1.690				14750						14750						1							
Gulley 9 – pipe	800			152.935	152.200	0.735	152.150	0.785	0.760	0.935		800				800					1				1								
Branch 1	2600			152.935	152.250	0.685	152.200	0.735	0.710	0.885		2600				0													2600		1	1	1
Branch 2	3000			152.935	152.250	0.685	152.250	0.685	0.685	0.860		3000				0													3000			1	1
Gulley 10 – pipe	800			152.805	152.160	0.645	152.110	0.695	0.670	0.845		800				800					1				1								
Branch 1	1400			152.805	152.210	0.595	152.160	0.645	0.620	0.795		1400				0													1400		1	1	1

Table 17.8 Pipework schedule 4.

DRAINAGE PIPE RUN: EXCAVATION SCHEDULE: NRM Part 4

1	2	3	4	5	6	7	8	9	10	11	12	13	14	15	16	17	18	19	20	21	22	23	24	25	26	27	28	29	30	31	32	33	34
				STAGE 1 DEPTHS							DEPTHS: 100mm dia.				DEPTHS: 150mm dia.						STAGE 2 PVC							STAGE 3 CI					
					HEAD		FOOT			PLUS BED 150mm PLUS SURROUND 25mm=	DEPTHS				DEPTHS							BEND		JUNCTION				DEPTHS				CI	
M/H	LENGTH			EGL	IL	DEPTH	IL	DEPTH	AVE		✓.50m	✓1.00m	✓1.50m	✓2.00m	✓.50m	✓1.00m	✓1.50m	✓1.00m	✓1.50m	✓2.00m	GULLEY	100	150	100	150	REDUCER	RE	✓500	✓1.00	✓1.50	BEND	CONN TO GULLEY	CONN TO WC
MH4 – RE	20000	19775		152.700	152.400	0.300	151.950	0.750	0.525	0.700		19775			0	19775	0										1						
Gulley 11 – pipe	800			152.700	152.175	0.525	152.125	0.575	0.550	0.725		800			0	800	0				1				1								
Branch 1	3000			152.700	152.225	0.475	152.175	0.525	0.500	0.675		3000			0		0												3000		1	1	1
Branch 2	3200			152.700	152.225	0.475	152.225	0.475	0.475	0.650		3200			0		0												3200			1	1
Branch 3	3000			152.700	152.225	0.475	152.225	0.475	0.475	0.650		3000			0		0												3000		1	1	1
Gulley 12 – pipe	800			152.700	152.400	0.300	152.350	0.350	0.325	0.500		800			0	800	0				1			1									
Branch 1	3000			152.700	152.450	0.250	152.400	0.300	0.275	0.450	3000				0	0	0											3000			1	1	1
Branch 2	3200			152.700	152.450	0.250	152.400	0.300	0.275	0.450	3200				0	0	0											3200				1	1
Branch 3	3000			152.700	152.450	0.250	152.400	0.300	0.275	0.450	3000				0	0	0											3000			1	1	1
	214900	50875	48275								19200	146925	31775	14750	800	58075	1600	12550	20975	14750	12	0	0	9	3	1	4	18400	39500	46000	20	34	34
34																																	34

Table 17.9 Pipework schedule 5.

DRAINAGE PIPE RUN: EXCAVATION SCHEDULE: NRM Part 5

1	2	3	4	5	6	7	8	9	10	11	12	13	14	15	16	17	18	19	20	21	22	23	24	25	26	27	28	29	30	31	32	33	34
							STAGE 1 DEPTHS								DEPTHS: 100mm dia.			DEPTHS: 150mm dia.			STAGE 2 PVC										STAGE 3 CI		
					HEAD		FOOT			PLUS BED 150mm PLUS SURROUND 25mm=		DEPTHS										BEND		PVC JUNCTION		REDUCER	RE		DEPTHS			CI	
M/H	LENGTH			EGL	IL	DEPTH	IL	DEPTH	AVE DEPTH		<.50m	<1.00m	<1.50m	<2.00m	.50m	<1.00m	<1.50m	<1.00m	<1.50m	<2.00m	GULLEY	100 / 150		100	150			<500	<1.00	<1.50	BEND	CONN TO GULLEY	CONN TO WC
	214900	50875	48275								19200	146925	31775	14750	800	58075	1600	12550	20975	14750	12			9	3	1	4	18400	39500	46000	20	34	34
	<500	19200												14750																46000			
	<1000	146925												31775																39500			
	<1500	31775												146925																18400			
	<2000	14750												19200																14750			
Say Manholes	2250													2250																20975.000			
5 450	**214900**													214900																12550			
																														1600			
																														58075			
																														800			
																														2250			
																														214900			

Table 17.10 Drainage quantities check.

QUANTITIES CHECK

Add up the total length of all the pipes as column 2

Reconcile with Stage 1 lengths as columns 12, 13, 14, 15

Reconcile with Stage 2&3 pipe runs as columns 16, 17, 18, 19, 20, 21, 29, 30, 31

Total length — Across manholes

	Column 2	214900

Total length — All Depths — Between manholes

Depth	Column	Value	Material	Size
<500	Column 12	19200	PVC	100mm Ø
<1000	Column 13	146925		
<1500	Column 14	31775		
<2000	Column 15	14750		
Across manholes	5 × 450	2250		
		214900		

Total length — All Depths — Between manholes

Depth	Column	Value	Material	Size
<500	Column 16	800	PVC	100mm Ø
<1000	Column 17	58075		
<1500	Column 18	1600		
<1000	Column 19	12550	PVC	150mm Ø
<1500	Column 20	20075		
<2000	Column 21	14750		
<500	Column 29	18400	CI	150mm Ø
<1000	Column 30	39500		
<1500	Column 31	46000		
Across manholes	5 × 450	2250		
		214900		

Table 17.3 Practical application: Drainage (*Continued*) 251

	.80	Drainage	12.55	Drain runs, ave 1.00 deep, uPVC 150mm dia., push fit, B & S granular [34.1.1.2
		Drain runs, ave 500 deep, uPVC 100mm diameter, push fit, B & S granular [34.1.1.1		
	58.08	Ditto, < 1.00, abd	20.98	Ditto, < 1.50, do
	1.60	Ditto, < 1.50, abd	14.75	Ditto, < 2.00, do
			18.40	Ditto, ave 1.00 deep, CI 100mm dia., mechanical seal joints, B & S concrete [34.1.1.1
			39.50	Ditto, < 1.50, do
			46.00	Ditto, < 2.00, do

Table 17.3 Practical application: Drainage (*Continued*)

		Drainage		The abbreviations 'abd' (as before described), 'ditto', and 'do' can be used to indicate repeated descriptions.
9/ 1		Fittings, uPVC, 100, junction [34.3.1.1		
3/ 3		Ditto, 150, do		*Provisional quantities have been used for the bends where none are shown on the drawings but some are likely to be used. They will be remeasured for the Final Account to include the accurate number that have been installed.*
1		Ditto, Reducer, 150/100, do		
				PROVISIONAL
			10	Fittings, uPVC, bend, 100. [34.3.1.1
12/ 1		Accessories, uPVC, gully, do [33.4.1.1	10	Ditto, 150, do.
			5	Ditto, Tee, 100, do
4/ 1		Ditto, RE, do	2	Ditto, 150
34/ 1		Connections, CI, to gully	5	Ditto, rest bend, 100
			2	Ditto, 150
34/ 3		Ditto, to WC	10	Raising pieces to gulley
			20	PCC lintels in brickwork walls

Table 17.3 Practical application: Drainage (*Continued*)

Drainage

	1	Manhole, size 600 x 840, 0.50–0.75 deep, PCC, 225mm thick base slab, 75mm thick wall, heavy duty cover slab, BS 5911, 700 x 990 x 150, bedded on mastic sealant, benching in-situ concrete, 1:6 [34.6.1.1	4/ 1		Access cover and frame, PCC, Grade C20, bedded and haunched in c/m on three courses 215mm thick Class B engineering brickwork, hole and cover [34.14.1.1
			2/ 1		Accessories, main channel, 100mm [34.4.1
			2/ 1		Ditto, 150, curved
			11/ 1		Branch channels, 100 [33.1.1
3/ 1		Ditto, size 690 x 1030, do, cover slab 840 x 1180 x150, do	4/2/ 1		PROVISIONAL Step irons [34.13.1.1
			Item		Testing & commissioning [34.17.1
		End of drainage			

17.4   SELF-ASSESSMENT EXERCISE: DRAINAGE TO PATIO AND DRIVE

Measure the drainage for the patio and drive using Drawing SDCO/2/E17/1 in Appendix 17. Please prepare your own measurement using schedules and blank double dimension paper in Appendix 1a and also prepare a query sheet of problems that you have encountered. Compare your own work with the proposed solution in Appendix 17 (Table E17.1) and self-assess your work on the assessment sheet in Appendix 1b.

To provide further assistance there are dedicated websites at http://ostrowski quantities.com and at Wiley Blackwell (http://www.wiley.com/go/ostrowski/ measurement). It is hoped that the provision of this will go some way towards explaining the concepts and principles more clearly than using the printed word alone.

18 Domestic Plumbing

18.1 Measurement information
- Drawings
- Specification
- Query sheet
- Insulation, sleevework and fire stopping

18.2 Technology

18.3 Practical application: Penthouse hot and cold water and soil and waste pipework

18.4 Self-assessment exercise: Two bed house

18.1 MEASUREMENT INFORMATION

Drawings

See Drawings SDCO/2/18/1 (hot and cold water plan), SDCO/2/18/2 (soil and waste pipework) and SDCO/2/18/3 (schematic cold water).

The drawings for plumbing, mechanical and electrical installations are diagrammatic. Full scale AutoCAD drawings in PDF format are on the website www.ostrowskiquantities.com.

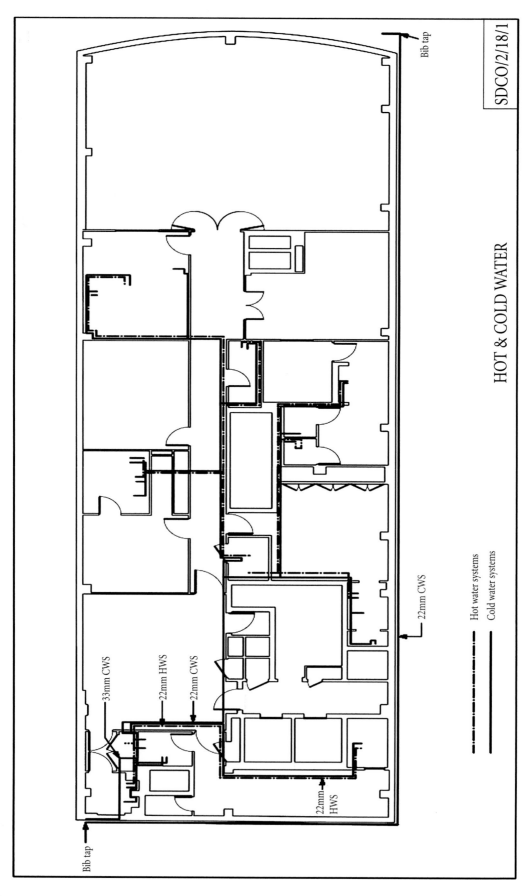

Drawing SDCO/2/18/1 Hot and cold water plan.

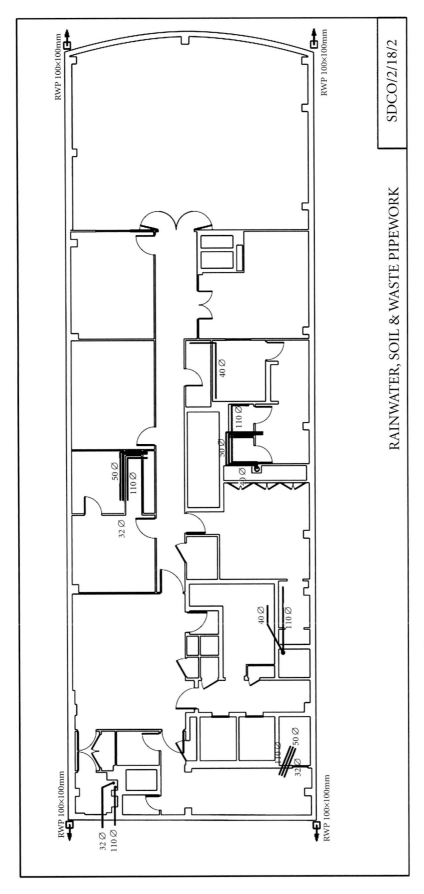

Drawing SDCO/2/18/2 Soil and waste pipework.

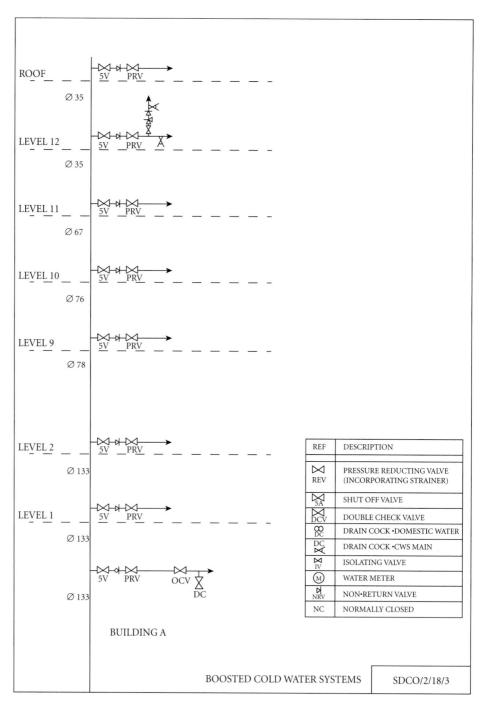

Drawing SDCO/2/18/3 Schematic cold water.

Specification

Table 18.1 Plumbing specification.

Specification S18 Plumbing
The above ground drainages and rainwater services shall be installed to the requirements of BS EN 12056, soil pipework to BS 4514, wastes to BS EN 15012, traps to BS EN 274. Copper pipework to BS EN 1057. Pressure reducing valves to be one size smaller than branch pipe size. Drain valves required to all low points on pipework. Cold water break tank in accordance with local bylaws. CWS installation to roof outlets to be insulated and trace heated. The overflow shall form an integral part of the cistern. Intumescent collar Terrain 'Firebreak'. Access doors to be fitted to ductwork. Pulsacoil cylinder by Gledhill ref. PCBP 150, 60 litre capacity.

Query sheet

Table 18.2 Plumbing query sheet.

QUERY SHEET	
Plumbing	
QUERY **(From the QS)**	**ANSWER** **(From the Architect/Engineer)** **(Assumptions are to be confirmed by QS)**
1. BWIC details assumed	Confirmed xx.xx.xx
2. Details of cold water break tank	
3. Access doors	Assumed, provisional
4. Sanitary ware	Assumed
5. Brassware for sanitary ware	Assumed

Insulation, sleevework and fire stopping

These items appear in several places in the NRM. They can be understood as follows in Table 18.3.

Table 18.3 Insulation, sleevework and fire stopping.

INSULATION, SLEEVEWORK AND FIRE STOPPING	
NRM REFERENCES	
INSULATION	
31 Insulation, fire stopping, fire protection	
31. 1–6 Insulation	All insulation except pipework and ductwork
38 Mechanical services	
38. 9–11 Insulation and fire protection to pipework	Insulation to pipework and ductwork.
38. 12–14 Insulation and fire protection to ductwork	This includes the plumbing pipework included in Section 33 Drainage above ground
SLEEVEWORK	
33 Drainage above ground	
33.4 Pipe sleeves through walls, floors and ceilings	Pipe sleeves for plumbing only
41 Builder's work in connection with mechanical, electrical & transport	
41.3 Pipe and duct sleeves	Pipe and duct sleeves for M & E installation
FIRE STOPPING	
31 Insulation, fire stopping, fire protection	
31.7 Fire stops, type stated	Fire sleeves and stops for plumbing only
31.8 Fire sleeves, collars and the like	
38 Mechanical services	
38.15 Fire stopping	Fire stopping for pipe and duct sleeves for M & E installation

18.2 TECHNOLOGY

The technology of plumbing is concerned with a large number of different kinds of fittings. The vocabulary is extensive. The measurement is simple. All pipe runs and insulation are measured as linear metres and everything else is enumerated. Each type and size of pipe is shown on the drawing and is measured separately. Start at the end of the largest and follow the pipe run from one end to the other and repeat until all runs are complete.

Diagram 18.1 provides representation of the typical plumbing installation and the names of the various fittings.

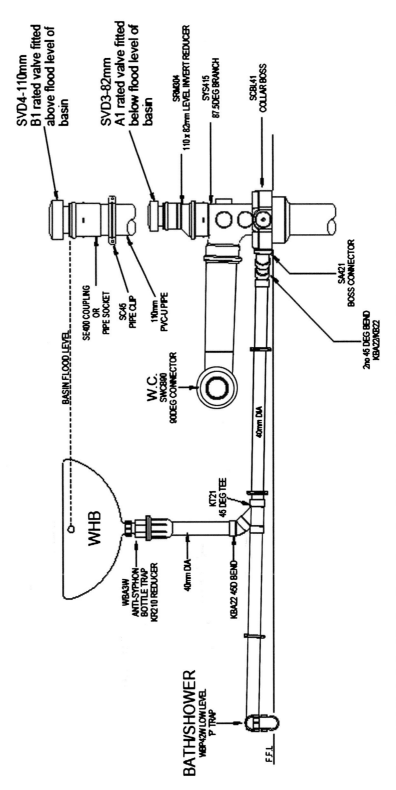

Diagram 18.1 Typical plumbing installation (reproduced with permission from Marley Soil and Waste Brochure 2012).

Table 18.4 Practical application: Penthouse hot and cold water and soil and waste pipework.

SERVICES

PLUMBING

Drawings

SDCO/2/18/1 H & C Water Plan

SDCO/2/18/2 Cold water schematic

SDCO/2/18/3 Soil and Waste Pipe Plan

Specification S18

The title page includes the trade name, the name of the contract, the full drawing schedule with dated revisions, the dated specification, the name of the measurer and the date.

SDCO
May 2012

Plumbing

To Take

<u>Sanitary ware</u>

<u>H & C water</u>
Cold water supply
Fittings and connections
Boilers
Valves
Trace heating
Hot water supply
Fittings and connections
Insulation

<u>Soil and waste</u>
Rainwater installation
Waste pipework
WC connections & traps
Ancillaries/fittings
Builders' work in connection (BWIC)

A 'to-take' list acts as a check list.

Plumbing comprises:

Sanitary ware in NRM 2 Section 32 Furniture, fittings and equipment.

Rainwater pipework in NRM 2 Section 33 Drainage above ground, rainwater installations.

Soil & waste pipework in NRM 2 Section 33 Drainage above ground, foul drainage installations.

Supply pipework in NRM 2 Section 38 Mechanical services.

BWIC in NRM 2 Section 41 BWIC.

			Plumbing					Each item of sanitary ware is usually a different specification and needs to be measured separately. They have been combined in this example for brevity.

Plumbing

Sanitary ware

Master bedroom
Wash hand basin 2
Bath 1
Shower 1
WC 1

Bed 2
Bath 1
WHB 1
WC 1

3/ 1

Bed 3
Bath 1
WHB 1
WC 1

WC
WC 1
WHB 1

Kitchenette
Sink 1
(Kitchen and kitchenette measured separately)

Cloakroom
WC 1
WHB 1

6/ 1

Totals
Baths 3
WHB 6
WC 5
Shower 1
Sink 1

Each item of sanitary ware is usually a different specification and needs to be measured separately. They have been combined in this example for brevity.

Ideal Standard, PO box 60, Kingston on Hull HU5 4J

Bath, 'Brasilia', 1794 x 796, complete with front and end panels, 'Jetline' pillar taps ref 6170, 50mm chromium plated waste, overflow, plug & chain, and all necessary fittings as manufacturers recommendations
[32.2.3

Wash hand basin, Ideal Standard, abd, 'Brasilia' WHB, ref 01.2 with pedestal, 'Jetline' monoblock mixer taps, ref 6070, 38mm chromium plated waste, overflow, plug and chain, complete

Table 18.4 Practical application: Penthouse hot and cold water and soil and waste pipework (*Continued*)

Plumbing

4/ 1		WC suite, Ideal Standard, abd, 'Brasilia' WHB, ref 02.2 with side outlet 'P' trap, 9 litre close coupled cistern, ref 884 and white plastic seat and wrap over cover		10/ 1		Extra over, connections to sanitary ware, < 65 mm [32.3.1
				4/ 1		Ditto, >65mm [32.3.2
1		Sink top, Leisure or equal and approved, stainless steel single drainer sink, size 1200 x 600, with 2 nr. Pillar taps, chromium plated waste, overflow, plug and chain, fixed to base unit (m.s.) to match kitchen units		Item		BWIC, sanitary ware [41.1

Plumbing

		Boosted cold water service (BCWS)		**Cold water service to boiler (CWS)**
			1	Pulsacoil ref. PCBP 150, 160 litres [38.1.1
4.00 / 1.00		Copper pipework, 35mm Ø header [Vertical [Horizontal [38.3.1		
			4.00 / 2/ .50	Copper, 28mm Ø [Horizontal [38.3.1
1		Ancillaries, pressure release valve, 35mm Ø [38.5.1	1	Ancillaries, I valve, 28mm Ø [38.3.1
1		Supply valve, ditto		
1		Tee ref. KT21, ditto [38.4.1	1	Tee ref. KT21, ditto
1		End, ditto	1	Stopcock, ditto
1		Stopcock, ditto	Item	BWIC, CWS to boiler [41.1
1		Reducer, 35/28mm Ø		
Item		BWIC, boosted CWS [41.1		*BWIC is measured as an item for each service as 41.1. Note 1. Individual items of BWIC are included at the end of this section.*

Table 18.4 Practical application: Penthouse hot and cold water and soil and waste pipework (*Continued*)

Table 18.4 Practical application: Penthouse hot and cold water and soil and waste pipework (*Continued*)

		Plumbing			Perimeter Cold water service
					Trace heating is a 'rogue' item. It is measured here because it is an integral part of the external pipework. It would be included in the Electrical BQ.
	12.00	Cold water (CWS)			
2/	4.00	Copper, 22mm Ø [38.3.1		12.00	Copper, 22mm Ø [38.3.1
	2.00			37.00	
	5.00	[Cloaks		1.00	&
	13.00	[Kitchenette			
	3.00	[Bed 2			Trace heating
	7.00				
2/	3.00	[Bed 3			&
	12.00				
3/	3.00				Insulation, 22mmØ, finished with PIB [38.9.2.1
	17.00	[Kitchen			
1.2/	3.00				
	5.00	[Cloak			
	5.00				
	3.00			1	Ancillaries, I valve, 28mm Ø [38.3.1
	18.00	[En suite			
2/	3.00				
2/	1	Ancillaries, Stopcock		1	Stopcock, ditto
				1	Bib tap
6/	1	Bends, ref KA450 [38.4.1		1	Cap for hot tub
	Item	BWIC, CWS [41.1		Item	BWIC, perimeter CWS

Plumbing

			9.00	Copper, 15mm Ø	
			5.00	[Kitchenette	
		Hot water services (HWS)	3.00	[38.3.1	
		Copper, 22mm Ø	5/ 1	Bends	
2.00		[38.3.1			
2.00		[Cloaks	1		
2.00		[Laundry		Tee	
8.00		[Bed 2			
3.00		[Bed 3/K	1	End	2.00
24.00					2.00
2/3.00		[WC			2.00
17.00		[Ensuite			8.00
					3.00
					24.00
					2/3.00
					17.00
					64.00
					9.00
1		Ancillaries, stopcock, 22mm Ø			5.00
					3.00
		[38.5.1			17.00
			64.00	Insulation, 22mmØ	
6/ 1		Bends, ref. KA450	17.00	Insulation, 15mmØ	
4/ 1		[38.4.1			
8/ 1					
3				Couplings, 35mm Ø	
9/ 1			2	[33.3.1	
				Ditto, 28mm Ø	
			2		
1				Ditto, 22mm Ø	
2/ 1			2		
1		Tee, ref. KT21		Ditto, 15mm Ø	
22/ 1			2		
				BWIC, HWS	
			Item	[41.1	

Table 18.4 Practical application: Penthouse hot and cold water and soil and waste pipework (*Continued*)

		Plumbing			Foul drainage installations (Soil & waste pipework)
		Rainwater			
4/ 3.00		Pipework, PVC, 100 x 100 , vertical [33.1.1	1.00 1.50 1.00 3.00 3.00		PVC, 110mm Ø [33.1.1 [Cloaks [K [B 3 [Ensuite
2/ 5.50 2/ 1.00		Ditto, horizontal	1.2/ 2.00		
2/2/ 1		Ancillaries, shoe [33.2.1	4.00		Ditto, at high level [33.1.1
		&			
		EO , 90° bend [33.3.1	8/4.00		Ditto, vertical [33.1.1
Item		BWIC, RW [41.1	4/ 1 1		Ancillaries, junction, 110mm Ø [33.2.1
			5/ 1		Extra over, manifolds, 2way [33.3.3.2
			4/ 1		Ditto, 3 way [33.3.2.3

Table 18.4 Practical application: Penthouse hot and cold water and soil and waste pipework (*Continued*)

Plumbing

Foul drainage installations   (Soil & waste pipework)

2/6.00 4.00 6.00 6.00	PVC, 50mm Ø waste [33.1.1 [ES [Kitchenette [B3 [ES	6.00 4.00 6.00	PVC, 32mm Ø waste [33.1.1 [Cloaks [Kitchenette [B3
		5/ 1	EO, bend ref KBA22 [33.3.1
4.00	Ditto, soil and vent pipe at high level [33.1.1	5/ 1	EO reducer 40/32, ref KR210
9 — 1	EO, bend, ref KBA22		
2/ 4.00	PVC, 40mm Ø waste [33.1.1		
2	EO, bend		

		Plumbing			
		Soil & waste pipework			*Many fittings will not be shown on the drawings but will be necessary. A selection of items will allow the work to be priced and included in the final account.*
5/	1	Extra over, WC connector, ref SWC890 [33.3.1			PROVISIONAL
3/	1	Ditto, traps, bath		2	Pipework ancillaries, straight WC connector, ref SWC890 [33.2.1
6/	1	Ditto, traps, ref WBA3W, WHB			&
	1	Ditto, traps, ref WBR42W, shower			Single branch
					&
2/	1	Ditto, traps, washing machine			Collar boss, ref SCBL41
					&
2/	1	Ditto, trap, sink			Access saddle
		Ditto, rodding eye access			&
5/	1				B1 valve, ref SVD4
		(PROVISIONAL)			&
		Extra over, chromium plating, 32mm Ø & Ditto, 40mm Ø & Ditto, 50mm Ø			A1 valve, ref SVD3
	1.00				&
					Boss connection, vertical, ref SA421
	Item	BWIC, soil & waste [41.1			

Table 18.4 Practical application: Penthouse hot and cold water and soil and waste pipework (*Continued*)

Plumbing

Pipe sleeves and fire stopping are important for safety, extensive and expensive but not shown on drawings. Allowances should be included for sleeves fixed during pipework installation, NRM 2 Section 33.4, and for fire stopping, NRM 2 Section 31.

		PROVISIONAL			PROVISIONAL
2/ 1		Pipe sleeves, copper pipework, 35 diameter, drylining, PVC sleeves, sealed with fire protection compound [33.4.1.1.1	2/ 1		Fire stops, sprayed proprietary fire protection compound, 300 x 300, vertical, including collar to pipe, 2hr fire rating [31.7.1
5/ 1		Ditto, 28 dia, do	2/ 1		Ditto, 200 x 200, do
10/ 1		Ditto, 22 dia, do	5/ 1		Ditto, 100 x 100, do
10/ 1		Ditto, 25 dia, do	5/ 1		Ditto, 100 x 100, do, 1 hr fire rating
3/ 1		Ditto, PVC pipework, 110 dia, do			
10/ 1		Ditto, 50 dia, do			
10/ 1		Ditto, 32 dia, do			

Plumbing

3/ 1			PROVISIONAL Access panels, Easy - Klix or equal and approved, 100 x 100, 1 hr fire rating [32.1.3	5/ 1		Identification, plates [33.9.1 & Ditto, discs & Ditto labels
				Item		Testing and commissioning, plumbing [33.10.1
				Item		O & M Manuals [33.12
				Item		Marking position of holes, mortices and chases in the structure, plumbing [41.2

End of plumbing

18.4   SELF-ASSESSMENT EXERCISE: TWO BED HOUSE

Measure the plumbing to a domestic house using the plan on Drawing SDCO/2/18/E1 in Appendix 18. Please prepare your own measurement using blank double dimension paper in Appendix 1a and also prepare a query sheet of problems that you have encountered. Compare your own work with the proposed solution included in Appendix 18 (Table E18.1) and self-assess your work on the assessment sheet in Appendix 1b.

To provide further assistance there are dedicated websites at http://ostrowski quantities.com and at Wiley Blackwell (http://www.wiley.com/go/ostrowski/measurement). It is hoped that the provision of this will go some way towards explaining the concepts and principles more clearly than using the printed word alone.

19 Mechanical Services

19.1 MEASUREMENT INFORMATION

Drawings

See Drawings SDCO/2/19/1 (air conditioning plan), SDCO/2/19/2 (air conditioning sections) and SDCO/2/19/3 (air conditioning details).

The drawings for plumbing, mechanical and electrical installations are diagrammatic. Full scale AutoCAD drawings in PDF format are on the website www.ostrowskiquantities.com.

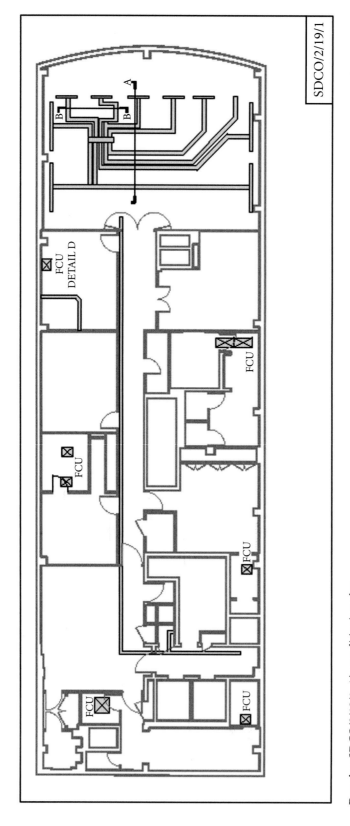

Drawing SDCO/2/19/1 Air conditioning plan.

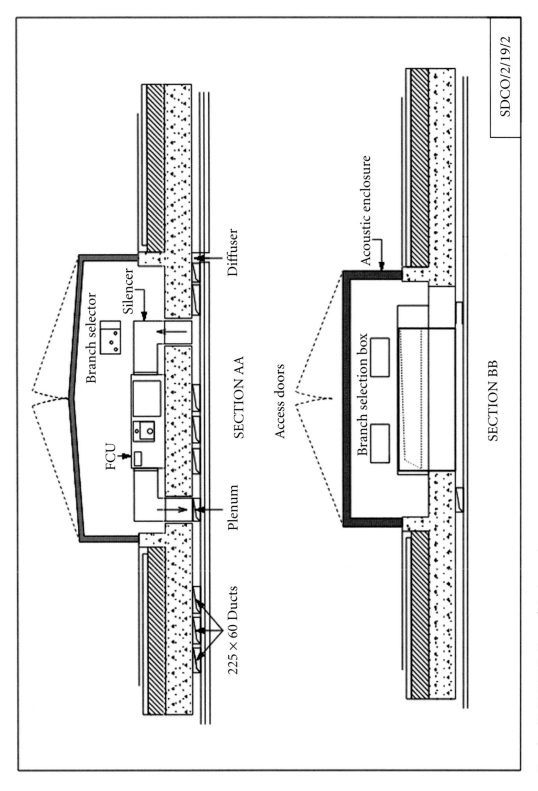

Drawing SDCO/2/19/2 Air conditioning sections.

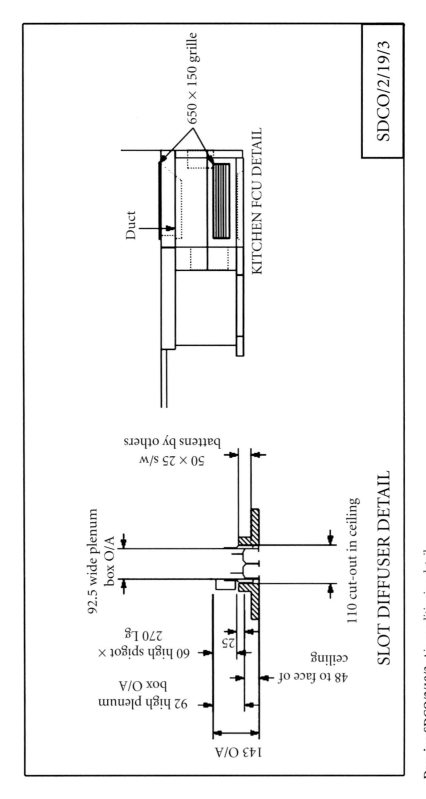

Drawing SDCO/2/19/3 Air conditioning details.

Specification

Table 19.1 Air conditioning specification.

Specification S19 Air conditioning
Localised variable air volume heat recovery system with separate condensers and primary plant on the roof in plant rooms and refrigerant pipework (measured separately).
Daikin fan coil units by Space Air Ltd or equal and approved for heating and comfort cooling.
FCU fixed to the structural ceiling in the suspended ceiling void. Separate supply and return grilles.
Access to the FCU to be a proprietary swing down sealed access panel sized to suit the size of the FCU.

Query sheet

Table 19.2 Air conditioning query sheet.

QUERY SHEET	
AIR CONDITIONING	
QUERY (From the QS)	ANSWER (From the Architect/Engineer) (Assumptions are to be confirmed by QS)
1. The FCU will not provide humidified air 2. Plenum boxes and branch selector units to direct and balance the air flow to be located outside the apartment 3. Zoning from the electrical controls separates daytime living room from night time bedrooms (measured separately)	Confirmed Confirmed

Insulation, sleevework and fire stopping

These items appear in several places in the NRM. Chapter 18 sets out the measurement requirements.

19.2   TECHNOLOGY: AIR CONDITIONING

The technology of mechanical services is complex but the measurement is simple. All pipe, duct and cable runs are measured as linear metres and everything else is enumerated. For insulation and fire protection to ducts the unit is square metres. Each type of duct is shown on the drawing and is measured separately. To measure, follow a duct run from one end to the other and repeat until all runs are complete. In this example the ductwork and fan coil units are measured, the flow and return and refrigerant pipework is not included.

Table 19.3 Practical application: Penthouse air conditioning.

SERVICES

AIR CONDITIONING

Drawings

SDCO/2/19/1 Plan

SDCO/2/19/2 Section

SDCO/2/19/3 Detail

SDCO/2/19/4 Section & detail

Specification S19

The title page includes the trade name, the name of the contract, the full drawing schedule with dated revisions, the dated specification, the name of the measurer and the date.

To Take
Fan coil units
Grilles
Access panels
Diffusers
Plenum
Ceiling hatches
The following are not measurable
Connections
Reducers
Labours
Bends
Tees
Ends
Boxes

SDCO
May 2012

Table 19.3 Practical application: Penthouse air conditioning (*Continued*)

		Comfort cooling			Connections to grilles are not measurable as 38. 6.*.*.*.1
	1	Terminal equipment and fittings, Daikin fan coil unit, in ceiling void, Type FXYSP32KA7V19, mechanically fixed to soffit of reinforced concrete slab above [38.2.1.1.1	2/ 1		Terminal equipment and fittings, grilles, 840 x 213, finished to Architect's details, fixed to high level FCU [38.2.1.1.1
6/ 1		Ditto, FXYSP25K7V1, do	2/ 1		Ditto, 990 x 213, do
				1	Ditto, 1230 x 213, do
	1	Ditto, FXYSP125KA7V19, roof mounted	2/ 1		Ditto, 1640 x 213, do
				1	Ditto, 2000 x 150, do
	1	Ditto, FXN25LVE , vertical chassis, complete with ductwork and grilles as Detail D [Kitchen	4/ 1		Ditto, access panels, 800 x 650, do
6/ 1		Ditto, branch selector unit, do		1	Ditto, 1050 x 750, do
				1	Ditto, 1600 x 650, do

Table 19.3 Practical application: Penthouse air conditioning (*Continued*)

			Comfort cooling				Ventilation supply ducts, galvanised sheet metal, HVCA DW 142, section 7, grade Z2, class C, rectangular section, 225 x 50, jointing between sheets, fixing at high level in ceiling voids, to soffit of structural reinforced concrete slab.
3/ 1			Terminal equipment and fittings, Gilbert, extruded aluminium alloy, linear slot diffuser, single deflection, Type GDS 2, polyester organic powder coated, standard colours, 1053mm long, finished to Architect's details, screw fixed to high level FCU [38.2.1.1.1	14.00 1.00 6.50 22.00			[38.6.1.1.1 [Lift [Supply
2/ 1			Ditto, 1189mm, do	3.00 7.00 4.00			[Plenum
1			Ditto, 1510mm, do	1.2/6.00	75.50		
4/ 1			Ditto, 2500mm, do				Ditto, 225 x 60, do [Lounge
4/ 1			Ditto, access panels, stool mounted, fixed with cam fasteners, including handles, 800 x 650, do	5.00 11.00 10.00 9.00 7.00 3.00 4.00			
1			Ditto, 1050 x 750, do	4.00 4.00			
1			Ditto, 1600 x 650, do	5.00	62.00		*Bends, angles, tees, etc. are significant labour items but are not measurable in the NRM*

Table 19.3 Practical application: Penthouse air conditioning (*Continued*)

			Comfort cooling			
			2/<u>225</u> 450			
			2/<u>50</u> <u>100</u>			
			<u>550</u>	Item	Identification, plates [38.16.1	
			2/<u>225</u> 450			
			2/<u>60</u> <u>120</u>			
			<u>570</u>			
				Item	Ditto, discs	
75.50 <u>.55</u>	41.53		Insulation, >25mm thick, 225 x 50, conductivity <0.04W/m °C, to high level ducts (Provisional) [38.12.1.1	Item	Ditto, labels	
				Item	Testing [38.17	
62.00 <u>.57</u>	35.34		Ditto, 225 x 60, do (Provisional)	Item	Commissioning [38.18	
<u>10</u>			Duct sleeves, 250 x 75, drylining, 1 hr FR (Provisional) [41.3.1.1.1	Item	O & M manuals [38.20	
<u>10</u>			Ditto, 250 x 100, do (Provisional)	Item	BWIC, mechanical services [41.1.1	
<u>10</u>			Fire stopping, 250 x 75, drylining, fire protection compound (Provisional) [38.15.1	Item	Marking position of holes, etc [41.1.2	
<u>10</u>			Ditto, 250 x 100, do (Provisional)		*Measurement of insulation, sleeve work and fire stopping are set out in Ch. 18 Plumbing.*	
			End of comfort cooling			

19.4   SELF-ASSESSMENT EXERCISE: VENTILATION

Measure the ventilation ducting using the plan on Drawing SDCO/2/19/E1 in Appendix 19. Please prepare your own measurement using blank double dimension paper in Appendix 1a and also prepare a query sheet of problems that you have encountered. Compare your own work with the proposed solution in Appendix 19 (Table E19.1) and self-assess your work on the assessment sheet in Appendix 1b.

To provide further assistance there are dedicated websites at http://ostrowski quantities.com and at Wiley Blackwell (http://www.wiley.com/go/ostrowski/ measurement). It is hoped that the provision of this will go some way towards explaining the concepts and principles more clearly than using the printed word alone.

20 Electrical Services

20.1 Measurement information
- Drawings
- Specification
- Query sheet
20.2 Technology
20.3 Practical application: Penthouse small power
20.4 Self-assessment exercise: Lighting

20.1   MEASUREMENT INFORMATION

Drawings

See Drawings SDCO/2/20/1 (small power plan) and SDCO/2/20/2 (power symbols).

The drawings for plumbing, mechanical and electrical installations are diagrammatic. Full scale AutoCAD drawings in PDF format are on the website www.ostrowskiquantities. com.

Measurement Using the New Rules of Measurement, First Edition. Sean D.C. Ostrowski.
© 2013 John Wiley & Sons, Ltd. Published 2013 by John Wiley & Sons, Ltd.

Drawing SDCO/2/20/1 Small power plan.

SDCO/2/20/1

SDCO/2/20/2

SHAVER UNIT

HEATED TOWEL RAIL

FLEX OUTLET PLATE

G   GRILL
H   HOB
OV   OVEN
TR   TOWEL RAIL

HL   HIGH LEVEL
LL   LOW LEVEL

ANALOGUE ADDRESSABLE
FIRE ALARM INTERFACE UNIT

ANALOGUE ADDRESSABLE
IONISATION SMOKE DETECTOR

ANALOGUE ADDRESSABLE
OPTICAL SMOKE DETECTOR

FAP   FIRE ALARM PANEL

PSU   24V POWER SUPPLY UNIT

VIDEO ENTRY PHONE UNIT

THREE PHASE CONSUMER
DISTRIBUTION BOARD

8   BATHROOM EXTRACT FAN

KITCHEN EXTRACT HOOD

20A DOUBLE POLE ISOLATOR
WH   WATER HEATER

2 GANG, 13A SWITCHED SOCKET OUTLET
450mm FFL

FUTURE 13A SOCKET OUTLET POSITION
(ALLOW 1000mm LOOP OF CABLE FOR
FUTURE CONNECTIONS)

13A SWITCHED FUSED CONNECTION UNIT,
WITH NEON AND FLEX OUTLET

FCU   FAN COIL UNIT
TR   TOWEL RAIL
MB   MOTORISED BLIND (FUTURE POSITION)
P   WATER HEATER PUMP/CYLINDER CONTROL
BS   A/C BRANCH SELECTOR
EM   ESPRESSO MACHINE
MC   MICRO-COMBI
EF   EXTRACTAFAN
SP   SHOWER PUMP

FUTURE 13A SPUR POSITION (ALLOW
1000mm LOOP OF CABLE FOR FUTURE
CONNECTIONS)

13A UNSWITCHED FUSED CONNECTION UNIT,
WITH NEON AND FLEX OUTLET
FP   FIRE ALARM PANEL
PU   POWER SUPPLY UNIT

1 GANG, 13A UNSWITCHED SOCKET OUTLET
NON-STANDARD 450mm FFL

5 AMP 1 GANG SWITCHED SOCKET
OUTLET

Drawing SDCO/2/20/2 Power symbols.

Specification

Table 20.1 Small power specification.

Specification S20 Small power
All electrical cables are to be PVC insulated, PVC sheathed incorporating an earth protective conductor with copper conductors to comply with BS 7211 and BS 7671and BASEC approved.
The small power wiring to be single core 600/1000V grade PVC insulated stranded copper conductor of 2.5mm² minimum cross section, to BS 6231, where practical.
The consumer unit is to be from the 'NS' range as manufactured by Wylex Limited, Wylex Works, Sharston Road, Wythenshawe, Manchester, M22 4RA (Tel: 0161 998 5454) or similar. Preference is for the surface fixed thermoplastic insulated units, except where recessing in a wall is necessary when a metal case may have to be considered. The unit should be supplied with 'NSB' type miniature circuit breakers and a Split Load Residual Current Device. The appropriate number of ways should be provided with a spare way on each side of the split load. Each MCB is to be clearly and permanently labelled to identify the circuit it is controlling. All but the lighting circuit and smoke detector circuit should be protected by the RCD.
All fittings are to be by MEM 250 or similar. Forbes and Lomax fittings as indicated on the drawings.

Query sheet

Table 20.2 Small power query sheet.

QUERY SHEET	
SMALL POWER	
QUERY (From the QS)	**ANSWER** (From the Architect/Engineer) (Assumptions are to be confirmed by QS)
1. Allowance of 5m per socket for small power. 2. Allowance of 20m × 2 for return = 40m per circuit for lighting.	Confirmed Confirmed

20.2 TECHNOLOGY

The technology of electrical services is complex but the measurement is simple. All cables are measured as linear metres and everything else is enumerated. The descriptions come from the specification supplied by the services engineer. To measure follow a cable run from one fitting to the next and repeat until the circuit is returned to the fuse box.

Table 20.3 Practical application: Penthouse small power.

			SERVICES SMALL POWER Drawings SDCO/2/20/1 Plan SDCO/2/20/2 Cable schematic Specification S20 SDCO May 2012				*The title page includes the trade name, the name of the contract, the full drawing schedule with dated revisions, the dated specification, the name of the measurer and the date.*

Table 20.3 Practical application: Penthouse small power (*Continued*)

			Small power		3/ 1		Ditto, shower unit
49/	1		Terminal equipment, switched socket outlet, 2 gang [39.2.1.1.1		2/ 1		Ditto, optical smoke detector
39/	1		Ditto, SSO, 2G, brass, Forbes and Lomax ref 123		4/ 1		Ditto, smoke alarm
6.54/	1		Ditto, 13A fused connection unit		2/ 1		Ditto, connection box for door magnet supply
3/	1		Ditto, 5A, 1G, SSO		1		Ditto, 3 phase consumer unit
4/	1						
Say	50.00		Ditto, 20A, double pole isolator		5		Fire sleeves, Terrain 'Firebreak', or equal and approved, 50 dia (Provisional) [38.15.1
			Ditto, cable to 20A double pole isolator		5		Ditto, 1 hr. (Provisional)
4/	1		Ditto, towel rail complete with cable and connection		5		Ditto, 100 dia, do (Provisional)

Table 20.3 Practical application: Penthouse small power (*Continued*)

		Small power				Cable length is an allowance of 5m from switch to fitting. An alternative is an allowance for each circuit.
		39				
		49				
		3				
		4				The drawings indicate additional allowances for future connections.
		95				
		2 gang 95 x 5m = 475				
		Future connections				NRM 2 states that all measurement is net. Clause 3.3.2(1)(a), p.46. This means there are no allowances for waste. Electricians allow for tails for final connections to all points when ordering cable but this is not a measurable item.
		29 x 1m = 29				
		Others 64 x 5m = 320				
		Perimeter Say 100				
		924				
924.00		Cables, XPLE PVC sheathed 90c copper to CMA code 6181e for internal wiring, 2.5mm twin and earth, clipped to background [39.5.1				
50.00		Cable containment, plastic, in walls (PROVISIONAL) [39. 5.1.1				Cables will be above the suspended ceiling, below the raised floor and within the plasterboard partitioning. The use of conduit will be minimal. A provisional allowance has been included.
Item		Identification, plates [39.14.1				
Item		Ditto, discs [39.14.2		Item		BWIC, small power services [41.1.1
Item		Ditto, labels [39.14.3		Item		Marking position of holes, etc. [41.1.2
Item		Testing [39.15				Cutting mortice for electrical switch boxes, plasterboard partitions, 50 x 50 (Provisional) [41.9.1.1
Item		Commissioning [39.16		10		
Item		O & M manuals [39.18		**End of Electrical work**		

20.4   SELF-ASSESSMENT EXERCISE: LIGHTING

Measure the lighting using Drawings SDCO/2/20/E1 and SDCO/2/20/E3 in Appendix 20. Please prepare your own measurement using blank double dimension paper in Appendix 1a and also prepare a query sheet of problems that you have encountered. Compare your own work with the proposed solution in Appendix 20 (Table E20.1) and self-assess your work on the assessment sheet in Appendix 1b.

To provide further assistance there are dedicated websites at http://ostrowski quantities.com and at Wiley Blackwell (http://www.wiley.com/go/ostrowski/measurement). It is hoped that the provision of this will go some way towards explaining the concepts and principles more clearly than using the printed word alone.

21 External Works

Measurement Using the New Rules of Measurement, First Edition. Sean D.C. Ostrowski.
© 2013 John Wiley & Sons, Ltd. Published 2013 by John Wiley & Sons, Ltd.

21.1   MEASUREMENT INFORMATION

Drawings

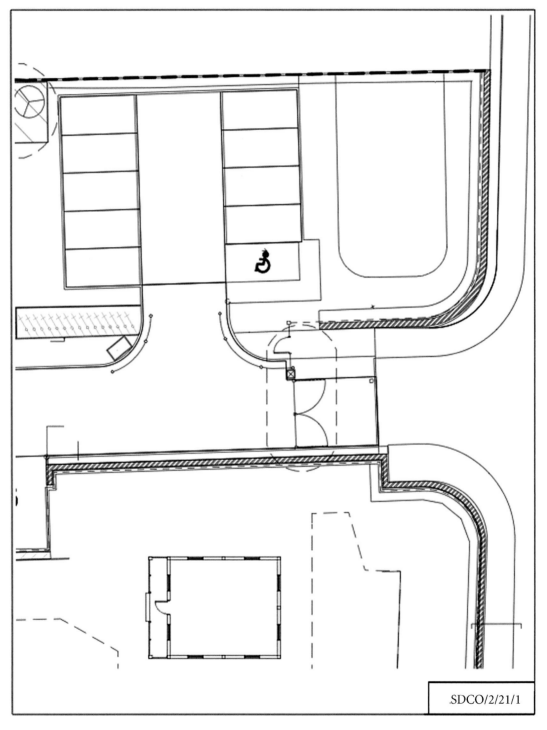

.SDCO/2/21/1

Drawing SDCO/2/21/1 Retaining wall plan 1.

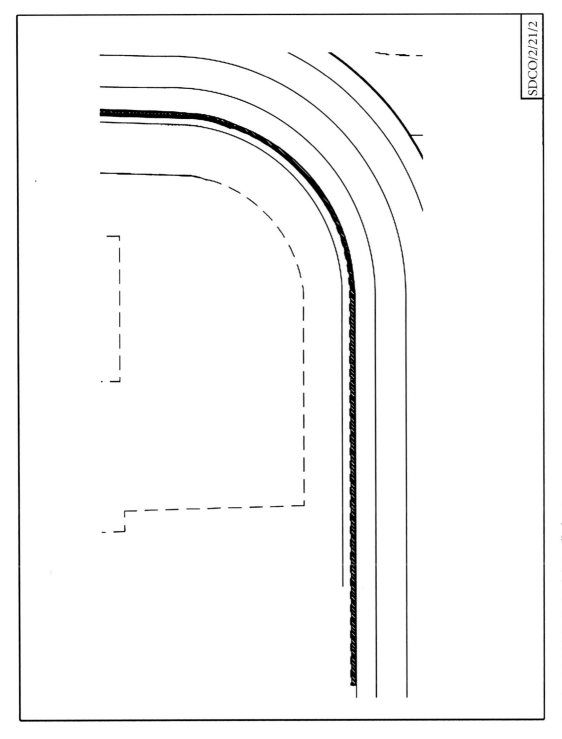

SDCO/2/21/2

Drawing SDCO/2/21/2 Retaining wall plan 2.

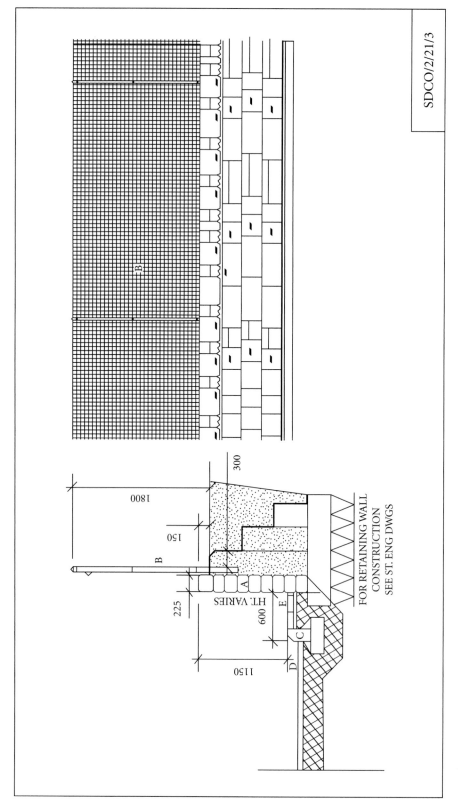

Drawing SDCO/2/21/3 Retaining wall elevation and section.

SDCO/2/21/3

300

1800

150

B

225

HT. VARIES

A

600

E

C

D

1150

FOR RETAINING WALL
CONSTRUCTION
SEE ST. ENG DWGS

Specification

Table 21.1 Retaining wall specification.

Specification S21 Retaining wall & Fencing
All surplus excavated material to be removed from site.
Imported hardcore as fill to be broken brick and the like blinded with 50mm sand.
Mass concrete foundations and backfill BS12, designed mix, grade C20, 40mm aggregate.
Masonry Lias limestone to match existing, 225 thick, 150mm × random length and height. Gauged mortar1:2:9. Bond English Garden wall bond. Jointing to be brush finished to coarse texture.
Coping, hen and cock.
Steel paladin fencing, 1800 high, RHS posts 60 × 60 × 2mm at 2975 centres, galvanised wire, >40gm², horizontal wires 3mm diameter at 50mm centres, with additional "V" beams using 4mm wiring, vertical wire 3mm diameter at 50mm and 12.5mm centres, welded joints, nylon coated, polyester organic powder coated to BS 6496, >100 microns, fixed to posts with spot welding and mechanical stainless steel screws.
Entrance gates, height 1800 at the hinge, 2100 at the centre × 5500 wide overall, comprising mild steel vertical bar gate, BS 1722-9, type 9, verticals to be 20mm diameter at 145 centres, 50 × 50 SHS frame, inserted through holes in the horizontal section and welded at the bottom, mechanical fixing with fish plates, cleats and bolts to BS 4190, complete with hinges 50 × 75 × 10mm mild steel cleats and 15mm diameter × 75mm bar welded to cleat, power installation.

Query sheet

Table 21.2 Retaining wall query sheet.

QUERY SHEET	
EXTERNAL WORKS	
QUERY (From the QS)	ANSWER (From the Architect/Engineer) (Assumptions are to be confirmed by QS)
1. Spoil to be removed from site 2. Scale of Section 3. Working space 4. Backfill to be Type 1 fill 5. Assume English Garden wall bond to match existing bond	Confirmed xx.xx.11 Use figured dimensions. Confirmed xx.xx.xx Nil. Confirmed xx.x.xx Yes. Confirmed xx.xx.xx

21.2 TECHNOLOGY

There is a slope across the site. The foundations to the retaining wall to the north of the gates require fill using high quality Type 1 fill. To the south of the gate is the cut area and the foundation requires a trench. The wall is built from the stonework that is available on the site. The measurement of the wall takes no account of the masonry available on site. The extent of new masonry is at the discretion of the contractor and is adjusted in the pricing of the work. The existing precontract water level is unknown.

Table 21.3 Practical application: Retaining walls.

			External works					
			RETAINING WALL					
			HOSTEL					
			Drawings					
			SDCO/2/21/1 Plan					
			SDCO/2/21/2 Plan					
			SDCO/2/21/3 Plan/Elevation					
			SDCO/2/21/4 Plan/Elevation					
			Specification S20					*The title page includes the trade name, the name of the contract, the full drawing schedule with dated revisions, the dated specification, the name of the measurer and the date.*
			SDCO					
			March 2012					

Table 21.3 Practical application: Retaining walls (*Continued*) 299

External works

EXCAVATION

	Average EGL South		Average EGL North
Dwg 1	57120	Dwg 1	55500
	57110		55180
	57500		53360
	57590		55140
	56990		53420
	56320		272600
	56710		÷ 5
	55680	Ave EGL =	54520
Dwg 2	54040		
	55520		
	57150		
	621730		
	÷ 11		
Ave EGL =	56520		

Level to top of retaining wall   56900

	Wall	1150		
Less	coping	(150)	1000	
	Paving	50		
	Base	150		
	Footing	250		
	Hardcore	250	700	(1700)

FORMATION LEVEL   55200

South Wall EGL	56520		North Wall EGL	54520
FL	(55200)		FL	(55200)
DIG	1320		FILL	(680)

Table 21.3 Practical application: Retaining walls (*Continued*)

Retaining wall

EXCAVATION

Length of walls (All scaled)

	South 34500	
	16000	
	27000	77500
Curved	(2 x π x 15000)/4	23565
Curved	(2 x π x 4000)/4	6283
	North	16500
Curved	(2 x π x 5000)/4	7550

77.50 1.30 1.32	Foundation excavation, ne 2m deep, trenches [South [5.6.2.1		2/77.50 1.32	Earthwork support, not exceeding 2m deep. (PROVISIONAL) [5.8.1
23.57 1.30 1.32 7.86 1.30 1.32	Ditto, curved		2/23.57 1.32 2/ 6.28 1.32	Ditto, curved (PROVISIONAL)
			35.00 1.30 .68	Imported fill, to base of trench, Type 1, >500 deep [5.12.3
Item	Disposal, off site [5.9.2		7.86 1.30 .68	Ditto, curved, do

Earthwork support is not a measurable item in the NRM unless specifically required by the Contract Administrator NRM 2 Ref 5.8.1.1.1.1. In this case the site is sloping and some of the trenches are curved. A provisional quantity has been measured.

Curved work is deemed to be included as NRM 2 p.13, Note 6. However for clarity it is measured separately as NRM item 5.6.2.*.*.*.2

Table 21.3 Practical application: Retaining walls (*Continued*)

		Retaining wall				
		EXCAVATION				Formwork is necessary to the fill area of the foundations because there is no trench
77.50 1.30		Imported filling, sand blinding bed, ne 50mm thick, 50mm thick, level, retaining wall	16.50			Plain formwork, sides of foundations, 500mm high retaining wall
23.57 1.30		[5.12.1.1.1				[11.13.1
6.28 1.30		___m² x 0.05 = ___m³				
16.50 1.30		&	8.55			Ditto, curved, radius 9000, do
7.86 1.30		Imported filling, Type 1, ne 500mm thick, 250mm thick, do.				1000 77500 700 23565 400 6284 2100 16500 7855 131704
		[5.12.2.1.1				
		___m² x 0.25 = ___m³				
		&				
		Mass concrete, any thickness, in trench filling, poured against earth	131.70 .30 2.10			Vertical work, mass concrete, Grade C20, <300mm thick, retaining walls
		[11.1.1.2.1				[11.5.1.1
		___m² x 0.25 = ___m³				
			½/131.70 1.00 1.00			Sundry in-situ concrete work, backfill, >300 wide, horizontal, retaining walls
						[11.6.2.1

Table 21.3 Practical application: Retaining walls (*Continued*)

Retaining wall

CONCRETE

2/ 77.50	1.00	Plain formwork, faces of retaining wall, vertical [11.22.1
2/ 16.50	1.00	
77.50	.70	Ditto, to single side, do [11.22.*.*.1
16.50	.70	
77.50	.40	
16.50	.40	

The following work does not appear in the NRM. However curved formwork is a significant cost item and should be measured.

2/23.57	1.00	Ditto, curved, 15.00m radius
23.57	.70	Ditto, curved to 15.00m radius, single side
23.57	.40	

2/ 6.28	1.00	Plain formwork, faces of retaining wall, vertical, curved, 4.00m radius
6.28	.70	Ditto, to single side, do
6.28	.40	
2/ 7.86	1.00	Ditto, curved, 5.00m radius
7.86	.70	Ditto, to single side, do
7.86	1.00	

These additional labours to the formwork are measured because they are complex items with substantial costs.

77.50	16.50	Extra over, arris, 50 x 50mm.
23.57		Ditto, curved, 15m
6.28		Ditto, 4.00 radius
7.86		Ditto, 5.00m radius

Table 21.3 Practical application: Retaining walls (*Continued*)

Retaining wall

77.50 1.45 16.50 1.45	Walls, Lias limestone to match existing, 225 thick, built against retaining wall, 150mm x random length and height, in g/m 1:2:9, English Garden wall bond, joints to be finished with coarse texture [14.1.4.4.1.1		77.50 16.50	Ditto, bands, horizontal, coping, cock and hen to match existing, do [14.7.1.3.4
		1150 2/150 500 1450	23.57	Ditto, curved, 15.00m radius, do [14.7.4.4
23.57 1.45	Ditto, curved to 15.00m radius, do [14.1.4.4.6		6.28	Ditto, curved, 4.00m radius, do
			7.86	Ditto, curved, 5.00m radius, do
6.28 1.45	Ditto, curved, 4.00m radius, do			
7.86 1.45	Ditto, curved, 5.00m radius, do			

End of Retaining wall

21.4   SELF-ASSESSMENT EXERCISE: FENCING

Measure the fencing using the plan and section on Drawings SDCO/1/21/1, SDCO/1/21/2 and SDCO/1/21/3 in this chapter. Please prepare your own measurement using blank double dimension paper in Appendix 1a and also prepare a query sheet of problems that you have encountered. Compare your own work with the proposed solution in Appendix 21 (Table E21.1) and self-assess your work on the assessment sheet in Appendix 1b.

To provide further assistance there are dedicated websites at http://ostrowski quantities.com and at Wiley Blackwell (http://www.wiley.com/go/ostrowski/ measurement). It is hoped that the provision of this will go some way towards explaining the concepts and principles more clearly than using the printed word alone.

22 Preliminaries

22.1 INTRODUCTION

Preliminaries have been examined in the companion volume *Estimating and Cost Planning Using the New Rules of Measurement*. They are also included in this volume to provide a comprehensive review of all the elements that require measurement. Like the quantities for the building works the preliminaries require accurate and consistent measurement. NRM 1 introduces a comprehensive schedule of costs to be included and many items also have a method of measurement. The preliminaries are shown in NRM 1 Part 4 Element 9 pp. 277–306 and NRM 2 Section 2.7 pp. Section 1 pp. 50–119. Although the NRM indicates what should be measured there are different methods of measurements for different types of preliminaries. The general rule is that measurement provides more accuracy than a percentage and that quotations from specialists are better than educated guesses. A detailed spreadsheet of each item is required to measure preliminaries. An example is set out below.

Measurement Using the New Rules of Measurement, First Edition. Sean D.C. Ostrowski.
© 2013 John Wiley & Sons, Ltd. Published 2013 by John Wiley & Sons, Ltd.

Table 22.1 Sample spreadsheet layout for measuring preliminaries.

FUNCTION	SIZE		DURATION		RATE	TOTAL
Accommodation						
Site Manager	$20m^2$	×	100 wks	×	£3/m^2	6,000
Meeting Room	$50m^2$	×	100 wks	×	£3/m^2	15,000
Alterations/Repairs						
					Allowance	10,000
Cleaning	2hrs	×	100 × 5 days	×	£10/hr	10,000
Consumables			100 wks	×	£100/wk	10,000

An extract from the comprehensive schedule set out for Management and staff Element 10.2 NRM 1 p. 218 is shown in Table 22.2.

Table 22.2 Preliminaries example (NRM 2 p. 77).

Element 9.2: Main contractor's cost items; Item 1.2.1: Management and staff			
Component	**Included**	**Unit**	**Excluded**
1 Project specific management and staff	Main contractor's project specific management and staff such as: 1 Project manager/director. 2 Construction manager. 3 etc.	Week (number of staff by number of man hours per week by number of weeks)	1 Security staff (included in sub-element 1.2.4: Security)
etc.	etc.	etc.	etc.

There are several areas that require some additional information to improve the extent of measurement and accuracy. Some items are set out below.

Samples

A considerable proportion of work on site is now prefabricated. Samples are therefore necessary to ensure compatibility at the interface between trades and standards of workmanship. The relevant subcontractors should provide a quotation for the provision of samples during the design stage and immediately prior to commencement on site.

Commissioning

Commissioning is carried out by the most expensive labour on site, usually in restricted spaces with pressing deadlines to achieve. Costs need to include premium rates, hotel costs, travel expenditure and bonuses. Again, quotations from the subcontractors give a more accurate price.

Commissioning fuel will be required during the latter part of the contract for 24 hours per day and 7 days per week. Fuel consumption for each building can be provided by the services engineer.

Consumables/Expenses

Management contracts can include large numbers of off-site meetings which generate considerable expenditure for travel and subsistence.

Scaffolding

The cost of scaffolding is usually a quotation from a subcontractor. However the cost of adaptations and additional hire can increase the cost substantially. Sufficient allowances for adaptations and additional hire should be included for each part of the scaffolding quote.

Small plant

Battery driven tools and hand held tools are often provided in standard proprietary containers from specialist plant hire companies. Popular tools often run out and replacements are very expensive. Again, suitable allowances for replacements should be added to the quotation.

Security

Security on site usually becomes ineffective after 18 months due to complacency of the subcontractor staff who are often casual agency personnel. Allowances should be included in the costs for regular replacement of security companies.

The use of swipe card systems is quite commonplace. However storage and retrieval of the data is not usually included in the quotation. Suitable allowances should be included in the preliminaries for adequate record keeping and retrieval.

Cleaning

Allowances in the preliminaries are usually for a 'builder's' clean prior to handover. However the client representatives usually require a higher level of cleanliness, the 'sparkle' clean standard. Again, suitable allowance should be included in the preliminaries for several cycles of cleaning.

Snagging

Work to be remedied during the defects liability period should be carried out by the relevant subcontractor. However this is often multitask work and requires addition work from a specialist snagging team. This requires further allowances in the preliminaries.

The period just before and just after practical completion can mean the damage of protective covers as they are repeatedly removed and replaced. Further allowances for additional protection are required in the preliminaries.

Attendances

Most subcontracts require additional attendances from the main contractor to clean and maintain work areas and the site. Further allowances for additional attendance are required in the preliminaries.

Lost and stolen

Although insurance is an overhead, lost and stolen items may come below the excess threshold of most all-risk construction insurance policies. This is particularly the case for final fix electrical and sanitary ware after the trade handover and before the practical completion. Allowances for direct replacement costs should be included in the preliminaries.

Access

Changes to the access requirements cannot usually be anticipated nor calculated. However allowances for these changes should be included.

Hoarding

Site hoarding is now extensively used for advertising purposes. The cost of the hoarding can be offset against advertising revenue or can be undertaken directly by the advertising company.

Catering

On site catering is subsidised by the main contractor to ensure breaks and meals are procured without excessive loss of time caused by going off site for refreshments.

Traffic management schemes

Traffic management schemes require temporary traffic lights for 24 hours per day and 7 days per week. Fuel consumption for a petrol generator requires a man and a van to service the generator. The cost of this servicing far outweighs the hire cost of the equipment.

Calculations of preliminaries and temporary works

Suggestions on how to calculate these figures are included in Table 22.3.

Table 22.3 Preliminaries calculations required: 1.

CALCULATION OF PRELIMINARIES AND TEMPORARY WORKS

DESCRIPTION			INFORMATION			CALCULATION	
PRELIMINARIES							
STANDARD TERMS							
	Tendering alterations		Ajusting errors/adding new info.			Estimate of cost/day	
	Increased costs		NEDO/BCIS/Estimate			Extrapolate indices.	
	Fees		D.S. Fees			Published Tables/P. Sum	
			Health & safety			Estimate/Provisional sum	
	Samples		Preambles			Subcontractor quote	
			Assembly			Estimate	
	Commissiong fuel		Assess time period			Estimate	
	Taxes		Landfill tax			Percentage	Provisional sum
			VAT			New work	0%
						Brownfield	5%
						Refurbishment	17.5%
	Insurances		Contractors All Risk			Percentage	
			Employers liabilities			Percentage	
			Joint refurbishment			Percentage	

(Continued)

Table 22.3 (*Continued*)

CALCULATION OF PRELIMINARIES AND TEMPORARY WORKS

DESCRIPTION	INFORMATION	CALCULATION
MANAGEMENT		
Detailed schedule of personel x time period		Estimate
Number of meetings		Estimate
Extent of travel		Estimate
Hotels/expenses		Estimate
PLANT		
Detailed schedule or each item x time period		Quote
Alterations and adjustments		Estimate
Small site plant		Estimate
CONSUMABLES		
Detailed schedule or each item x time period		Estimate
EXPENSES		
Detailed schedule or each item x time period		Estimate
SECURITY		
Agree specification with architect.		Quote
HANDOVER/ COMMISSIONING		
Preambles for procedures. Specialist gangs x time period.		Estimate
Premium rates/additional hours/hotels/travel		Estimate

Table 22.4 Preliminaries calculations required: 2.

CALCULATION OF PRELIMINARIES AND TEMPORARY WORKS

DESCRIPTION		INFORMATION		CALCULATION	
SUBCONTRACTORS					
	Remedials/snagging			Included in quote	
	Attendances/clearing up. Number of gangs x time period			Estimate	
WARRANTIES					
	Performance Bond	Percentage of contract sum		Quote	Provisional sum
	Collateral warranties	Percentage of contract sum		Quote	Provisional sum
	Subcontractor warranties	Percentage of subcontract sums		Quote	Provisional sum
RISK REGISTER					
	Schedule of at risk items.	Assessed percentage risk. Cost of each delay.			Provisional sum
OVERHEADS					
		Particular to the contract		Percentage	
		Regional		Percentage	
		Group/HQ.		Percentage	
PROFIT					
		Published accounts/Directors review			
		Strength of order book		Percentage	

Table 22.5 Preliminaries calculations required: 3.

CALCULATION OF PRELIMINARIES AND TEMPORARY WORKS

DESCRIPTION	INFORMATION	CALCULATION
TEMPORARY WORKS		
ACESS		
Roads	Hardcore and removal	Measure for Rate/m²
Parking	Tarmac and removal	Measure for Rate/m²
Unloading	Heavy duty h/c & removal	Measure for Rate/m²
Alterations		Estimate for lump sum
FENCING		
	Hoarding	Measure for Rate/m
SITE HUTTING		
Prepare detailed schedule of accommodation for purchase or rent		Measure the m² for quote
Heating & power		Service provider quote
Furniture and fittings	Purchase or rent from suppliers schedule or rates	
	Short term rental or reuse existing	
Telephones/Power/water	Can be calculated per month	Quote from suppliers
Catering/subsidies		Quote from suppliers

Table 22.5 (*Continued*)

SCAFFOLDING			Quote from suppliers
	Adaptations	Cost per lift x No.of changes	Estimate
	Skips		Estimate
SECURITY		Lighting and alarms	Quote from suppliers
SERVICES		Diversions	Service provider quotes
UNDERPINNING/BUTRESSES		Specification from consultant	Quote from subcontractor
PROTECTION OF WORK		To be included in subcontractor quote	
	After Practical Completion	Meaure surfaces per m^2	Estimate
TRAFFIC CONTROL		Agree spec'n with Architect	Quote from subcontractor
DEWATERING		Assessment of plant/mth	Estimate

22.2 OVERHEADS AND PROFIT (NRM 1 P. 39)

Overheads can accrue at local, divisional and international levels and all should be included in the costs for the particular site. However if the cost of overheads is not completely covered in the site specific tender it needs to be recovered elsewhere. As with plant, marginal costing can defer the cost of all the overheads to the head office. This is particularly the case with site financing charges which may be deferred to other sites or later time periods.

Profit levels are available from published accounts where profit on turnover can be ascertained. However the amount of profit to be included in a particular tender is dependent on the strength of the order book and the directors' view of the market.

Contracts that are project management, design and build or cost plus have subcontractor packages that include preliminaries, overheads and profit that accrue to the subcontractor. The management package will also include preliminaries and profit but is unlikely to reflect the full extent of either. Several of these packages will have preliminary allowances that are for the total contract.

It is unlikely that the analysis of the preliminaries, overheads and profit will be provided by the subcontractor, who may have no contractual relationship with the client.

22.3 PRACTICAL APPLICATION: SCHEDULE OF PRELIMINARIES

A typical example of site based preliminaries is set out in Table 22.6.

Table 22.6 Practical application: Preliminaries site wide.

PRELIMINARIES Site wide period: 30 weeks					Date:	2005
	Description	Qty	Unit	Rate	£	£
PLANT	Safety Inspections	6	Nr	250	1,500	
	Road Sweeper	26	Weeks	600	15,600	
	Sundry Plant	1	Item	3,000	3,000	20,100
ACCOMODATION	Hoarding (move and erect)	180	m	60	10,800	
	Hutting 300m^2 x £17.5/m2/p.a.	26	Weeks	101	2,625	
	Erect		Item		10,000	
	Dismantle		Item		5,000	
	Office Cleaning	26	Weeks	175	4,550	
	Computer Equipment	26	Weeks	75	1,950	
	Photocopier	26	Weeks	50	1,300	
	First Aid Equipment & Dressings	1	Item	500	500	
	Toilet Consumables	26	Weeks	25	650	37,375
SERVICES	Telephones	26	Weeks	200	5,200	
	Temporary Electrics	26	Weeks	200	5,200	
	Connection Charge for Temp. Water	1	Item	2,050	2,050	
	Water fountain	26	Weeks	30	780	13,230
STAFF	Producion Manager	26	Weeks	1,750	45,500	
	Site Manager - Shell & Core	26	Weeks	1,300	33,800	
	Site Manager - Fit Out	26	Weeks	1,300	33,800	
	Assistant Site Manager	26	Weeks	1,150	29,900	
	Planner 80%	26	Weeks	1,400	36,400	
	Typist/Site Clerk	26	Weeks	650	16,900	196,300
SKIPS	Hire of skips	78	Nr	115	8,970	8,970
SCAFFOLDING	Adaptions only				10,000	10,000
PROTECTION	Units	–	Nr	Nil	Nil	Nil
LABOUR	Street Cleaner	26	Weeks	475	12,350	
	Site Cleaning Labourer	26	Weeks	475	12,350	

(Continued)

Table 22.6 (*Continued*)

PRELIMINARIES Site wide period: 30 weeks		Qty	Unit	Rate	Date: £	2005 £
	Description	Qty	Unit	Rate	£	£
	Welfare Labourer	26	Weeks	475	12,350	
	Welfare Labourer	26	Weeks	475	12,350	
	Agency Supervisor	15	Visits	525	7,875	
	Handyman (Sales Centre)	26	Visits	750	19,500	
	Handyman	26	Weeks	750	19,500	96,275
SMALL TOOLS	Small Tools	26	Weeks	50	1,300	1,300
CONSUMABLES	Protective Clothing	10	Nr	150	1,500	
	Fire precautions	1	No	1,200	1,200	
	Progress Photographs	6	No	250	1,500	
	Consumables Stationery	26	Weeks	50	1,300	5,500
SECURITY	Security Gate (day)	26	Weeks	800	20,800	
	Security Gate (night)	26	Weeks	800	20,800	
	Induction	20	Weeks	800	16,000	
	Security Central Traffic Area	20	Weeks	800	16,000	73,600
SIGNAGE	Notices / Signs	1	Item	3,500	3,500	3,500
	SITE MANAGEMENT BUDGET TOTAL				466,150	466,150

22.4   SELF-ASSESSMENT EXERCISE: WEEKLY COSTS

Calculate the weekly cost for the 30 week contract period.

Compare your own work with the proposed solution in Appendix 22.

Self-assess your work on the assessment sheet in Appendix 1b.

To provide further assistance there are dedicated websites at http://ostrowski quantities.com and at Wiley Blackwell (http://www.wiley.com/go/ostrowski/ measurement). It is hoped that the provision of this will go some way towards explaining the concepts and principles more clearly than using the printed word alone.

23 Computer Aided Taking Off

23.1 INTRODUCTION

The automatic measurement of quantities direct from drawings remains an enticing but as yet unrealised goal. The development of AutoCAD and building information modelling (BIM) and a range of software measurement packages that provide digitisers, a standard range of specification descriptions, preformed phraseology for the particular method of measurement and automatic sorting into the correct sequence ensure that measurement procedures are in a compliant format. The technical competence required to measure the work is still required. This skill is to know what to measure and how to find it on the drawings, and this has not been replaced by software. However, much of the procedure of physical measurement can be replaced by screen based processes. Being able to understand and validate the result of computer aided taking off is now much more important. User error and mistakes made to input cannot be rectified without a full knowledge of the requirements of the NRM. How to use them now requires the QS to acquire these additional competencies and to be able to interrogate and correct them.

23.2 SCREEN BASED LEARNING

The pedagogy of learning to use a new piece of software is a process which requires practice, contingent instruction and contemporaneous feedback. Practice is best facilitated by following an example and repeating it with variations. Contingent instruction is

---

Measurement Using the New Rules of Measurement, First Edition. Sean D.C. Ostrowski.
© 2013 John Wiley & Sons, Ltd. Published 2013 by John Wiley & Sons, Ltd.

the process whereby questions are dealt with as they occur so that development can continue. Contemporaneous feedback enables corrections and improvement to be made with a minimum of trial and error.

This chapter uses computer aided taking off (CATO) to provide examples of how to measure using screen based software. It is a well known measurement package provided by Causeway Technologies Limited, originally developed by EC Harris. It is an excellent example of how the printed word has been superseded by screen based material. CATO provides a full screen based NRM library and the QS has to learn how to use it for selecting the appropriate descriptions and inserting the correct quantities. Like Word, Excel and Powerpoint it requires a substantial amount of practice to be able to use it. My own experience in teaching has indicated that a hard copy 'click by click' guide is extremely useful. In particular it restricts the extensive movement between different screen images to be able to find out what to do next. The practical example set out below is derived from the screen based NRM library in CATO. A short video clip is also available and demonstrates the procedure for selecting descriptions, inserting quantities and the publication of the BQ for the first item in the click by click guide, the lifting of turf. This is on the author's dedicated companion website http://ostrowskiquantities.com.

23.3   PRACTICAL APPLICATION: CLICK BY CLICK GUIDE TO CATO

Table 23.1 Click by click guide to CATO: 1.

SCREEN/GO TO	ACTIONS
ENTRY PROCEDURES	
Software v16.1.0 Log in	Causeway CATO suite User name: Administrator Password: XXXXXX
File	New Project
Reference Title Description	SDCO1 Foundations Substructure exercise 1 OK
Classification/Project classification Select project class/Standard Enter word(s) . . . Results	Click on the button with three dots '. . .' Ignore Select Education. Click OK
Go to Icon 'Take off & Bills'	Click
Project details	✓
Add new project	Yes
Take Off	Double click
DIM File	SDO (3 characters only)
Description	Foundations
	✓
This should give you the Measurement & Description screen.	
Procedures for making corrections and clearing and going back to the beginning, etc. are set out at the end of this section.	

Table 23.2 Click by click guide to CATO: 2.

SCREEN/GO TO	ACTIONS
The following guide provides a measurement for excavation on dimension sheets 1–6. Corrections can be implemented using the procedures on the first page.	
SIMPLE DESCRIPTIONS AND DIMENSIONS	
DIMENSION SHEET No. 000001	
To commence measurement with lifting the turf	
Go to Description	
Copy/Change	Double click
Level 1 Excavation and filling	Double click
Level 2 Site clearance/preparation	Double click
Level 3 Site preparation	Double click
Spec'n	✓
Level 4 Lifting turf for preservation	Double click
Level 5 Method and location of preservation stated	Double click
Select extended Level 5 phrase	Double click
By hand	✓
On site	✓
Watering	✓
	✓ & ✓
Description	✓
Cursor goes automatically to Dimensions	
9.40	Enter
20.40	Enter
Timesing column	Enter
At this point there is a complete description and dimensions in accordance with NRM2.	

Table 23.3 Click by click guide to CATO: 3.

SCREEN/GO TO	ACTIONS
DIMENSION SHEET No. 000002	
This sheet demonstrates excavating the top soil	
Stay on Dim Sheet 00001	
Go to Copy/Change	Click
This creates another dim sheet 000002	
Drag the cursor to the level 3 description	Click
Site preparation	Double click
Spec'n	✓
Level 4 Remove topsoil; depth stated	Double click
Extended Level 4 phrase	Double click
Enter 100	Double click
	✓ & ✓
Description	✓
Cursor automatically goes to the Dimensions	
9.40	Enter
20.40	Enter
Timesing column	Enter
REVIEW	Dim sheet 1◄
	Dim sheet 2►
At this point there is a complete description and dimensions in accordance with NRM2.	

Table 23.4 Click by click guide to CATO: 4.

SCREEN/GO TO	ACTIONS
DIMENSION SHEET 000003	
This sheet demonstrates Disposal of excavated material off site	
Stay on the same screen	
This is the third page of dimensions	
Description	Copy/Change
This creates another dim sheet 000003	
Drag the cursor to the level 3 description box	
Disposal	Double click
Spec'n	✓
Level 4 Excavated material off site	Double click
	✓ & ✓
Description	✓
Cursor automatically goes to Dimensions	
9.40	Enter
20.40	Enter
0.10	Enter
Timesing column	Enter
Review	Dim sheet 1◄
	Dim sheet 2◄
	Dim sheet 3►
At this point there is a complete description and dimensions in accordance with NRM2.	

Table 23.5 Click by click guide to CATO: 5.

SCREEN/GO TO	ACTIONS
DIMENSION SHEET 000004	
This sheet demonstrates Excavating pits	
Stay on the same screen	
This is the fourth page of dimensions	
Description	Copy/Change
This creates another Dim Sheet 000004	
Level 3 Drag the cursor to the description box	
Excavation, commencing level stated if not original ground level	Double click
Spec'n	✓
Level 4 Foundation excavation	Double click
Level 5 Not exceeding 2m deep	Double click
Select extended Level 5 phrase	✓
Pits	✓
	✓ & ✓
Description	✓
Dimension column	
1.25	Enter
1.25	Enter
1.70	Enter
Timesing column	
7	Enter
Review	Dim sheet 1/2/3◄
	Dim sheet 4►

(Continued)

Table 23.5 (*Continued*)

SCREEN/GO TO	ACTIONS
DIMENSION SHEET 000005	
Filling the pits. Stay on the same screen. This is the fifth page of dimensions	
Description	Copy/Change
This creates another dim sheet	
Level 3 Drag the cursor to the description box	
Filling obtained from excavated material	Double click
Spec'n	✓
Level 4 Final thickness of filling exceeding 500mm deep	Double click
Level 5 Source, distance, destination and method stated	✓
Select extended Level 5 phrase	✓
On site	✓
500m	✓
On site	✓
Mechanically	✓
Description	✓
Cursor auto automatically goes to Dimension	✓ & ✓
1.25	Enter
1.25	Enter
1.70	Enter
Timesing column	
7	Enter
Review	Dim sheet 1/2/3/4◄
	Dim sheet 5►

Table 23.6 Click by click guide to CATO: 6.

SCREEN/GO TO	ACTIONS
DIMENSION SHEET 000006	
Earthwork support. Stay on the same screen. This is the sixth page of dimensions	
Description	Copy/Change
This creates another dim sheet	
Level 3 Drag the cursor to the description box	Click
Support to face(s) of excavation where not at the discretion of the contractor	Double click
Spec'n	✓
Level 4 Maximum depth stated	Double click
Select extended Level 4 phrase	Double click
2m	✓
Level 5 Location stated	Double click
Select extended Level 4 phrase	Double click
Pits	✓
	✓ & ✓
Description	✓
Come out of the automatic programme by pressing ESC	
Dimensions	
Go to Measurement	Click
Go to Multiply mode	Click
1.25	Enter
2	Enter
Go to More measure	Click
1.25	Enter
2	Enter
Go to Multiply mode	Click
1.70	Enter
7	Enter
Review	Dim sheet 1/2/3/4/5◀
	Dim sheet 6▶

Table 23.7 Click by click guide to CATO: 7.

This section sets out the procedures for making corrections and clearing and going back to the beginning, etc.	
ESCAPE	
Come out of the automatic programme by pressing ESC	
CORRECTIONS	
CLEARING A DIM SHEET	
Measurement	
Nil entire dim sheet	Click
Are you sure?	✓
DESCRIPTIONS	
STANDARD PHRASES	
Drag the cursor to the relevant description box	Double click
Spreading	Click
	X
Warning	Yes
	✓
INCOMPLETE OR WRONG DESCRIPTION	
Come out of the automatic programme	Press Esc
DIMENSIONS	
Enter dimension column	
Dimension to be altered	Click
Make changes	Enter
EXIT & RE-ENTRY	
File	Click
File again	Click
List of projects	
SDCO7	Double click
Take off	
Dimensions Title	
Foundations	Double click

Table 23.7 (*Continued*)

ANDING ON	Click
Copy dims	Click
First dim	Click
Last dim	Click
Copied dims are highlighted in blue	
SIGN POSTS	
At the bottom of the dim sheet	▶ Scroll right
Drag the cursor to the last dim	
F2	
Enter the description in the signpost text box	
Hidden column 1	

23.4 SELF-ASSESSMENT EXERCISE: CATO

Complete the following CATO exercise for the measurement of reinforced concrete to slabs and compare your own work with the proposed solution set out in Appendix 23 and demonstrated on the second CATO video clip on the companion website http://ostrowskiquantities.com/. Self-assess your work on the assessment sheet in Appendix 1b.

To provide further assistance there are dedicated websites at http://ostrowski quantities.com and at Wiley Blackwell (http://www.wiley.com/go/ostrowski/ measurement). It is hoped that the provision of this will go some way towards explaining the concepts and principles more clearly than using the printed word alone.

24 Preparation of Bills of Quantities

> 24.1 BQ preparation
> • Measurement, working-up and BQ preparation
> • Preliminaries, Prime Cost and Provisional Sums
> • Pricing
> 24.2 Practical application: Abstract and BQ for reinforced concrete frame
> 24.3 Self-assessment exercise: Abstract and BQ for substructure

24.1 BQ PREPARATION

Measurement, working-up and BQ preparation

The measurement of a single trade or NRM section is a Bill of Quantities. When several sections are measured together they are collectively known as Bills of Quantities. They may be sent out to tender as individual trades; as a work package, eg the Shell and Core work package for the structural work or a Fit Out Package for the internal works; or for the complete works. The preparation of BQs follows a procedure which emphasises the accuracy of each stage before moving on to the next stage.

- 'Take-off' all the measurement from the drawings. This is the measurement of all the work that is required by the NRM.
- Measure any other work that does not conform to any NRM descriptions. These items are described as 'rogue' items.
- Check the transfer of all the dimensions from the drawings.
- Check the 'side-casts' (also called waste/calculations).
- Check the transfer of the side-casts to the dimension columns.
- 'Square' all the dimensions. This is calculating the sum or product of all the dimensions as m^3, m^2, lm, nr, items or others.
- Check the squaring of the columns.

Measurement Using the New Rules of Measurement, First Edition. Sean D.C. Ostrowski.
© 2013 John Wiley & Sons, Ltd. Published 2013 by John Wiley & Sons, Ltd.

- Checking and altering the dimensions are done in red to distinguish the alterations from the original.
- 'Work-up' the dimensions. This is gathering together all the same NRM items that may have been measured separately in different parts of the take-off. Spreadsheets are ideal for this purpose.
- Calculate the sum of all the transfers for each item.
- Check the transfer of the dimensions to the working-up schedules.
- Check the arithmetic.
- Schedule the items in the BQ in compliance with the order in NRM 2.
- Amend the descriptions to conform precisely to the NRM.
- Amendments to descriptions are made in green to distinguish them from the originals.

The emphasis in working-up is on checking each stage because the majority of errors are transfer errors. These errors occur in both manuscript and screen versions of measurement. The measured BQ is complete, arithmetically correct and in compliance with the NRM. The example in the practical application below shows a typical abstract and BQ layout.

Preliminaries, Prime Cost and Provisional Sums

In addition to the measured work there are several elements that need including to provide a complete price for the works. The first is the preliminaries and overheads and profit which is included in a previous chapter.

Major items of work being designed and built by specialist contractors are inserted in the BQ as Prime Cost (PC) Sums (NRM 2 pp. 12 and 49). Allowances for the main contractor's profit, overheads and attendances are included as follows:

PC sum for mechanical services		£1,500,000.00
Main contractor's overheads (as a percentage) Say 10%		£150,000.00
Main contractor's profit (as a percentage)	5%	£ 75,000.00
Main contractor's special attendances (as a percentage)		£ 100,000.00
Main contactor's general attendances (as a percentage)		£50,000.00
		£1,875,000.00

Another element that needs to be included is where there is insufficient information available to complete the measurement. An allowance is included in the appropriate section of the BQ as Provisional Sums (PS). Where the items can be reasonably defined they are described as defined Provisional Sums. Reasonably defined items are those that can be included in the programme of works and in the contract documentation, eg the addition of coloured glass in the glazing (NRM 2 2.9.1–3). Items that cannot be measured and cannot be included in the programme are included as undefined PS sums, eg structural alterations to a refurbishment after a contract has commenced or a decorative feature to be decided by the architect at a later stage in the contract (NRM 2 2.9.4–5). Some items can be reasonably anticipated, although they are not shown on the drawings, and they are included as provisional quantities, eg break out obstructions in the ground or sacrificial plates for switches and sockets. These PS items can only be carried out by instructions from the contract administrator. The original PS sum is omitted and the new work measured and valued and included in the final account.

The BQ is now complete and comprises the preliminaries, measurement, PC and PS sums. The intention of this document is to provide the contractor's estimator with enough information to be able to price the document without having to measure any of the work. This complies with the two main aims of the NRM, consistency and accuracy. The consistency comes from a standard method of measurement that is common to all and the accuracy from a document that measures all the work that is necessary using a prescriptive method of measurement.

The quantities conform to the trade descriptions of the work and include all the labour, materials and plant necessary to undertake the work. The estimator has detailed knowledge of the labour, materials and plant incorporated into each unit of measurement and will also have a record of the prices for the work that have been provided in the past by the their own trade subcontractors. The contractor is therefore able to provide an accurate price for the work based on the actual cost of carrying out the work.

Pricing

Accurate pricing for BQs suffers from the same problem as accurate pricing for estimates. The problems are rehearsed in the companion volume *Estimating and Cost Planning Using the New Rules of Measurement* and repeated here. The perennial problem with estimates is that they are not accurate enough. Despite the vast amount of information that is available this problem remains. Some research has been carried out by an eminent cost engineer in the USA. Hackney (1992) refers to a Rand study in Chapter 53 concerning Quantified Effects of Management Decisions. He states that '*Estimates prepared by groups with a vested interest in having the projects approved were found to be associated with added cost growth. Cost growth for their project averaged 49%, compared with 22% for estimates prepared by estimators independent of the group sponsoring the project . . .*'

This confirms our impression that all estimates are problematic. He goes on to state the reasons why '*The primary reason for this difference appears to be that the average estimate prepared by project champions were less well defined than the average estimate prepared by independent estimators. . . . The difference may also reflect the tendency of sponsor groups to provide their estimators with optimistic assumptions.*'

Complex buildings require complex estimates and this is often in advance of sufficient information being made available. No matter how experienced the estimator, the ability to predict an accurate price is restricted to quality and quantity of the information available. As a consequence the anticipated costs should be expressed as a range rather than a single target figure.

24.2   PRACTICAL APPLICATION: ABSTRACT AND BQ FOR REINFORCED CONCRETE FRAME

See Tables 24.1 and 24.2.

Table 24.1 Practical application: Abstract of reinforced concrete frame in Chapter 8.

REINFORCED CONCRETE FAME

Page references are to Chapter 8 dimension sheets

RC 20/25 Slabs/beams <300 [m³]	Frmwk, soffit <300,<3m [m²]	Ditto, S & S attached bms, 3-4.5m, profiled, 4 edges, girth 1800, detail X [m²]	Recess (3nr) shallow gutter, 800 x 50, top of slab [m]	Rebar, BS 449, G500C,
64.51 ✓ p.5	77.28 ✓ p.9	106.92 ✓ p.12	6.60 ✓ p.11	25.000 ✓ p.15
5.26 ✓ p.5	Ditto, 300-450,do [m²]		6.60 ✓ p.11	Ditto, bent
69.77 ✓ p.9	77.28 ✓ p.9	Columns, regular (12) <3 [m²]	Holes, <500, 250 [nr]	2.000 ✓ p.15
Ditto, bays 26m [m²]	Ditto <300, 3-4.5m high [m²]	50.56 ✓ p.13	4 ✓ p.11	Ditto, links
45.54 ✓ p.6	141.98 ✓ p.12	50.56 ✓ p.13	4 ✓	2.000 ✓ p.15
Slabs/beams >300 [m³]	Ditto, S&S attached bms, do [m²]	Columns, regular (12) 3-4.5 hi [m²]	EO, to edge, Detail X, curved, 50 dia [m]	Rebar, BS EN 430 high yield, ms, straight, 12mm
8.96 ✓ p.5	96.32 ✓ p.9	57.60 ✓ p.13	59.40 ✓ p.12	5.00 ✓ p.15
21.43 ✓ p.7	61.60 ✓ p.10	57.60 ✓ p.13	59.40 ✓	Ditto, bent
30.39 ✓	10.56 ✓ p.10	Sloping top surface<15° [m²]	EO ff 300hi [m]	2.000 ✓ p.15
RC 30/37 Columns, >300 [m³]	168.48 ✓	199.32 ✓ p.8	59.40 ✓ p.16	Ditto, links
10.82 ✓ p.8	Ditto, S & S, 3-4.5m hi [m²]	199.32 ✓	59.40 ✓	2.000 ✓ p.15
Power floating [m²]	55.72 ✓ p.13	Holes, 50 x 50, 200 th slab, <3m hi	Holes, 100 x 100, 200 th slab, <3m hi	Holes, 50 x 50, 400 wide bm, <3m hi
182.16 ✓	55.72 ✓	4 ✓ p.14	4 ✓ p.14	4 ✓ p.14
182.16	Ditto, upsd bm, irregular, ave <250hi, do [m²]	Ditto, 400	Ditto, 400	Ditto, 3-4.5 hi
	6.00 ✓ p.10	4 ✓ p.14	4 ✓ p.14	4 ✓ p.14
	(3.88) ✓ p.10	Ditto, 250, 3-4.5	Ditto, 250, 3-4.5	
	2.12 ✓	4 ✓ p.14	4 ✓ p.14	
	Ditto, 250-500,do [m2]			
	3.88			
	3.88 ✓ p.10			

The ticks next to each quantity indicates that it has been correctly transferred from the dimensions.

The ticks against the summmary figure indicates they are arithmetically correct.

Table 24.2 Practical application: BQ of reinforced concrete frame in Chapter 8.

THE NEW BUILDING

REINFORCED CONCRETE FRAME

BILLS OF QUANTITIES

for

NAME DEVELOPMENTS LTD

Ostrowski & Associates

May 2012

1/1

(*Continued*)

Table 24.2 Practical application: BQ of reinforced concrete frame in Chapter 8 (*Continued*)

						BQs 1/2/3 RC Frame	
						£	p
	THE NEW BUILDING						
	IN-SITU CONCRETE WORKS						
	BILL No.1						
	In-situ concrete						
	Reinforced in-situ concrete, BS 5328, section 4, designated mix RC 20/25						
						The ticks indicate that	
A	Horizontal work, not exceeding 300 thick, slabs and beams	Code	70✓	m³		*the quantities have been correctly transferred from the*	
						abstract.	
B	Ditto, in bays, 26m²	Code	46✓	m³			
C	Ditto, over 300 thick	Code	30✓	m³			
						They are to be	
	Reinforced in-situ concrete, BS 5328, section 4, designated mix RC 30/37					*removed from the final document.*	
D	Vertical work, over 300 thick, columns	Code	11✓	m³			
	Surface finishes to in-situ concrete						
						The column headed	
E	Power floating	Code	182✓	m²		*'Code' is for the insertion of the client's cost code references.*	
						The suffix letter I is not used, as it can be confused with the number 1.	
			1/2		To Collection £		

Table 24.2 Practical application: BQ of reinforced concrete frame in Chapter 8 (*Continued*)

							BQs 1/2/3 RC Frame
	THE NEW BUILDING: IN-SITU CONCRETE WORKS						
	BILL No.2						
	Formwork, plain						
A	Soffits of horizontal work, less than 300 thick, propping less than 3m high	Code	77✓	m^2			
B	Ditto, 300–450 thick, do	Code	77✓	m^2			
C	Ditto, less than 300 thick, propping 3–4.5m	Code	142✓	m^2			
D	Sides and soffits of attached beams, rectangular, propping not exceeding 3m high	Code	168✓	m^2			
E	Ditto, propping 3–4.5m high	Code	56✓	m^2			
F	Ditto, profiled, 1800 girth, four edges	Code	107✓	m^2			
G	Sides of upstand beams, irregular, average 250 high, propping 3–4.5m high	Code	2✓	m^2			
H	Ditto, 250–500 high, do	Code	4✓	m^2			
J	Sides of isolated columns, 12 nr, propping less than 3m high	Code	51✓	m^2			
K	Ditto, 12 nr, 3–4.5m high	Code	58✓	m^2			
L	Sloping top surface, less than 15°	Code	199✓	m^2			
M	Recess, shallow gutter, 800 × 50, in top of slab	Code	7✓	m			
N	Holes, diameter less than 500, depth less than 250	Code	4✓	nr			
P	Extra over, curved inside edge, 50mm diameter	Code	59✓	m			
							The suffix letter O is
Q	Extra over, fair face, 300 high	Code	59✓	m			*not used, as it can be confused with the number 0.*
	2/1		To Collection £				

(*Continued*)

Table 24.2 Practical application: BQ of reinforced concrete frame in Chapter 8 (*Continued*)

					BQs 1/2/3 RC Frame	
	THE NEW BUILDING: IN-SITU CONCRETE WORKS					
	BILL No.2					
	Formwork, plain					
	PROVISIONAL					
A	Holes 50 × 50, horizontal, slab, 200mm thick, propping not exceeding 3m high	Code	4✓	nr		
B	Ditto, 400mm thick, do	Code	4✓	nr		
C	Ditto, 250mm thick, propping 3–4.5m high	Code	4✓	nr		
D	Ditto, 100 × 100, 200mm thick, ne 3m high	Code	4✓	nr		
E	Ditto, 400mm thick, do	Code	4✓	nr		
F	Ditto, 250mm thick, propping 3–4.5m high	Code	4✓	nr		
G	Ditto, 50 × 50mm, beam 400 thick, propping not exceeding 3m high	Code	4✓	nr		
H	Ditto, 50 × 50, propping 3–4.5m high	Code	4✓	nr		
	2/2			To Collection £		

Table 24.2 Practical application: BQ of reinforced concrete frame in Chapter 8 (*Continued*)

							£	p
						BQs 1/2/3		
						RC Frame		
	THE NEW BUILDING						£	p
	IN-SITU CONCRETE WORKS							
	BILL No.3							
	Reinforcement							
	Reinforcement, BS4449, grade 500C, hot rolled mild steel							
A	Bars, straight	Code	25✓	t				
B	Bars, bent	Code	2✓	t				
C	Bars, links	Code	2✓	t				
	Reinforcement, BS EN 1.430, high yield stainless steel							
D	Bars, straight	Code	5✓	t				
E	Bars, bent	Code	2✓	t				
F	Bars, links	Code	2✓	t				
			To Collection	£				
	THE NEW BUILDING							
	IN-SITU CONCRETE WORKS							
	BILL No.1/2/3							
	COLLECTION							
	Bill 1 In-situ concrete	1/2						
	Bill 2 Formwork	2/1						
	Bill 2 Formwork	2/2						
	Bill 3 Reinforcement	3/1						
		3/1	To Summary £					

24.3　SELF-ASSESSMENT EXERCISE: ABSTRACT AND BQ FOR SUBSTRUCTURE

Prepare the abstract and BQ for the substructure measurement in Chapter 4, Substructure. Please prepare your own work and also prepare a query sheet of problems that you have encountered. Compare your work with the proposed solution in Appendix 24 (Table E24.1) and self-assess your work on the assessment sheet in Appendix 1b.

To provide further assistance there are dedicated websites at http://ostrowski quantities.com and at Wiley Blackwell (http://www.wiley.com/go/ostrowski/ measurement). It is hoped that the provision of this will go some way towards explaining the concepts and principles more clearly than using the printed word alone.

References

Gagne, R. (ed) (2002) *The Conditions of Learning*. Austin, TX: Holt, Rinehart & Winston.

Hackney, J. W. (1992) *Control & Management of Capital Projects*, 2nd edn. New York: McGraw-Hill.

RICS (2007) *The RICS Code of Measuring Practice*, 6th edn. London: RICS Publishing.

RICS (2012a) *The RICS New Rules of Measurement NRM 1: Order of Cost Estimating and Cost Planning for Capital Works*, 2nd edn. London: RICS Publishing.

RICS (2012b) *The RICS New Rules of Measurement NRM 2: Detailed Measurement for Building Works*. London: RICS Publishing.

Wood, D. (2001) 'Scaffolding, contingent tutoring and computer supported learning', *International Journal of Artificial Intelligence in Education*, 12: 280–92.

Measurement Using the New Rules of Measurement, First Edition. Sean D.C. Ostrowski.
© 2013 John Wiley & Sons, Ltd. Published 2013 by John Wiley & Sons, Ltd.

Index

Measurement Using the New Rules of Measurement, First Edition. Sean D.C. Ostrowski.
© 2013 John Wiley & Sons, Ltd. Published 2013 by John Wiley & Sons, Ltd.

Also available from Wiley Blackwell

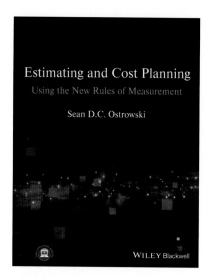

Estimating and Cost Planning
Using the New Rules of Measurement
Sean D.C. Ostrowski

9781118332658

www.wiley.com/go/construction